ADVANCES IN ASTRONOMY AND ASTROPHYSICS

SOLAR IRRADIANCE

TYPES AND APPLICATIONS

ADVANCES IN ASTRONOMY AND ASTROPHYSICS

Additional books and e-books in this series can be found on Nova's website under the Series tab.

ADVANCES IN ASTRONOMY AND ASTROPHYSICS

SOLAR IRRADIANCE

TYPES AND APPLICATIONS

DARYL M. WELSH
EDITOR

Copyright © 2020 by Nova Science Publishers, Inc.

All rights reserved. No part of this book may be reproduced, stored in a retrieval system or transmitted in any form or by any means: electronic, electrostatic, magnetic, tape, mechanical photocopying, recording or otherwise without the written permission of the Publisher.

We have partnered with Copyright Clearance Center to make it easy for you to obtain permissions to reuse content from this publication. Simply navigate to this publication's page on Nova's website and locate the "Get Permission" button below the title description. This button is linked directly to the title's permission page on copyright.com. Alternatively, you can visit copyright.com and search by title, ISBN, or ISSN.

For further questions about using the service on copyright.com, please contact:
Copyright Clearance Center
Phone: +1-(978) 750-8400   Fax: +1-(978) 750-4470   E-mail: info@copyright.com.

NOTICE TO THE READER

The Publisher has taken reasonable care in the preparation of this book, but makes no expressed or implied warranty of any kind and assumes no responsibility for any errors or omissions. No liability is assumed for incidental or consequential damages in connection with or arising out of information contained in this book. The Publisher shall not be liable for any special, consequential, or exemplary damages resulting, in whole or in part, from the readers' use of, or reliance upon, this material. Any parts of this book based on government reports are so indicated and copyright is claimed for those parts to the extent applicable to compilations of such works.

Independent verification should be sought for any data, advice or recommendations contained in this book. In addition, no responsibility is assumed by the Publisher for any injury and/or damage to persons or property arising from any methods, products, instructions, ideas or otherwise contained in this publication.

This publication is designed to provide accurate and authoritative information with regard to the subject matter covered herein. It is sold with the clear understanding that the Publisher is not engaged in rendering legal or any other professional services. If legal or any other expert assistance is required, the services of a competent person should be sought. FROM A DECLARATION OF PARTICIPANTS JOINTLY ADOPTED BY A COMMITTEE OF THE AMERICAN BAR ASSOCIATION AND A COMMITTEE OF PUBLISHERS.

Additional color graphics may be available in the e-book version of this book.

Library of Congress Cataloging-in-Publication Data

Names: Welsh, Daryl M., editor.
Title: Solar irradiance : types and applications / Daryl M. Welsh. Description: New York : Nova Science Publishers, [2020] | Series: Advances in astronomy and astrophysics | Includes bibliographical references and index. | Summary: "Solar Irradiance: Types and Applications first presents intelligent models for sizing, parameters forecasting and control of a photovoltaic system on the basis of a modified fuzzy neural net. The modified fuzzy neural net provides automatic fulfillment and modification of all proposed intelligent models. Following this, the authors discuss modeling direct normal irradiance at the Earth's surface. In addition to looking traditionally at direct normal irradiance as a fuel for concentrating solar systems, its use in computing the sunshine number is also explored. The closing study explores the potential of using simple empirical and artificial neural network models to estimate global solar radiation on a horizontal surface. Algeria was used as a case study and four statistical parameters were chosen to assess the performances of each model or approach"-- Provided by publisher.
Identifiers: LCCN 2020043982 (print) | LCCN 2020043983 (ebook) | ISBN 9781536187861 (paperback) | ISBN 9781536187991 (adobe pdf) Subjects: LCSH: Solar energy. | Solar radiation--Computer simulation. | Solar radiation--Measurement.
Classification: LCC TJ810 .S663 2020 (print) | LCC TJ810 (ebook) | DDC 621.47--dc23
LC record available at https://lccn.loc.gov/2020043982
LC ebook record available at https://lccn.loc.gov/2020043983

Published by Nova Science Publishers, Inc. † New York

CONTENTS

Preface		vii
Chapter 1	Photovoltaic Applications on the Basis of Modified Fuzzy Neural Net *Ekaterina A. Engel*	1
Chapter 2	Direct Normal Irradiance: Measurement, Modeling and Applications *Marius Paulescu*	81
Chapter 3	Potential Assessment of Global Solar Radiation Resources Based on Empirical and ANN Models: Algeria as a Case Study *T. E. Boukelia, M. S. Mecibah, A. Ghellab and A. Bouraoui*	119
Index		135

PREFACE

Solar Irradiance: Types and Applications first presents intelligent models for sizing, parameters forecasting and control of a photovoltaic system on the basis of a modified fuzzy neural net. The modified fuzzy neural net provides automatic fulfillment and modification of all proposed intelligent models.

Following this, the authors discuss modeling direct normal irradiance at the Earth's surface. In addition to looking traditionally at direct normal irradiance as a fuel for concentrating solar systems, its use in computing the sunshine number is also explored.

The closing study explores the potential of using simple empirical and artificial neural network models to estimate global solar radiation on a horizontal surface. Algeria was used as a case study and four statistical parameters were chosen to assess the performances of each model or approach.

Chapter 1 - The long-term contribution of solar energy is predicated on overcoming remaining technical barriers, mainly through research and development based on artificial intelligence. Data mining and machine learning methods gained a great attention in photovoltaic applications. The real-life photovoltaic systems have complex non-linear dynamic due to random variation of the system parameters and fluctuation of the solar irradiance. Thus, neural-network-based solutions have been proposed to approximate this complex dynamic. But the neuronet needs to become more adaptive. Modification of the network into a recurrent neuronet with fuzzy units provides intelligent behavior. This forms the motivation for the development of intelligent models for sizing, parameters forecasting and control of a photovoltaic system on the basis of a modified fuzzy neural net. Compared to existing fuzzy neuronets, including Adaptive Network-Based Fuzzy Inference System, the modified fuzzy neural net includes recurrent neuronets and fuzzy units. The authors generated the optimum architecture of the modified fuzzy neural net on the basis of development modified swarm intelligence algorithms. The function approximation capabilities of a neural net are exploited to approximate a membership function. This chapter presents the intelligent models for sizing, parameters forecasting and control of a photovoltaic system on the basis of a modified fuzzy neural net. The modified fuzzy neural net provides automatic fulfillment and modification of all proposed intelligent models. The first proposed model on the basis of a modified fuzzy neural net provides a two days ahead forecasting of the hourly power from a photovoltaic system under random perturbations. In order to train the effective modified fuzzy neural net the authors developed the modified multi-dimensional Particle Swarm Optimization, in which the multi-dimensional Particle Swarm Optimization is combined with the Levenberg-Marquardt algorithm. The comparison simulation results show that proposed modified multi-dimensional PSO algorithm outperforms multi-dimensional Particle Swarm Optimization and Levenberg-Marquardt algorithm in training the effective modified fuzzy neural net for a two days ahead forecasting of the hourly PV array power. The second proposed model on the basis of a modified fuzzy neural net is ambient temperature forecasting model under random perturbations. This model is important for photovoltaic systems. Simulation comparison results for a forecasting of ambient temperature demonstrate the effectiveness of the modified fuzzy neural net trained by proposed modified Ant Lion Optimizer as compared with the same one trained by Ant Lion Optimizer or Levenberg-Marquardt algorithm. The third proposed model on the basis of a modified fuzzy neural net is a photovoltaic system control model. According the photovoltaic system condition, the modified fuzzy neural net provides a maximum power point tracking under random perturbations. The simulation results demonstrate that the proposed photovoltaic system control model on the basis of a modified fuzzy neural net provides best performance and real-time control speed, as compared to a classical control model with a PID controller based on perturbation & observation, or incremental conductance algorithm. The fourth proposed model on the basis of a modified fuzzy neural net provides the best configuration and optimal sizing coefficient of a photovoltaic system. The simulation results demonstrate that the effectiveness of the proposed model is better than the Particle Swarm Optimization in sizing of a photovoltaic system. The fifth intelligent model on the basis of proposed modified Particle Swarm Optimization provides a maximum photovoltaic array power point tracking under fast-changing non-uniform solar irradiance level. Simulation comparison results demonstrate the competitive effectiveness of the proposed modified Particle Swarm Optimization as compared with the perturbation & observation algorithm and Particle Swarm Optimization in finding the global maximum power point of a partially shaded photovoltaic array.

Chapter 2 - The monitoring of solar radiation experienced a vast progress, not only through the expansion of the measurement networks, but also through the improvement of data quality. Nevertheless, the number of stations for monitoring direct normal irradiance (DNI) is still too small for achieving an accurate global coverage. Alternatively, various models for estimating DNI are exploited in many applications. This chapter deals with modeling DNI at the Earth's surface. It is structured in three sections. The first section introduces the methods of measuring DNI at the ground level. The second section summarizes the main classes of models for estimating DNI. The focus shifts upon the parametric class. Within this class, the atmospheric transmittance is explicitly expressed as a function of meteorological parameters. Choosing a highly-performant model is frequently limited by the availability of the parameters required for its running. Approaches for inferring the parametric models, along with the performances and weaknesses of the current models are reviewed. The third section is devoted to analyzing some applications of DNI. In addition to looking traditionally at DNI as a fuel for concentrating solar systems, its use in computing the sunshine number (a binary indicator stating whether the Sun is shining or not) is discussed.

Chapter 3 - Solar energy is considered to be a clean, safe, and economic source of energy, and has the potential to be the most exploited energy source in the future. The amount of this type of energy, which refers to solar radiation available on the earth's surface, depends on different specifications between physical and astronomical, meteorological and geographical ones. These parameters or specifications can be expressed by and not limited to: distance between earth and sun, extraterrestrial radiation, different astronomical and solar angles, atmospheric transmittance, as well as sunshine duration, ambient temperature, humidity and cloudiness at the relevant locations. To design and optimize any solar conversion system, the knowledge of accurate solar radiation data is extremely important. The best solar radiation data on the place of interest would be that measured at this specific location in a continuous and an accurate manner over a long term. However, for many developing countries, solar radiation measurements are not easily available due to financial or even technical limitations. On the other side, several spatial databases provide solar radiation values developed by different procedures, various spatial and temporal coverages, and different time intervals and space resolutions. However, these databases show different values of uncertainties due to the different approaches and the inputs used to generate them. Therefore, it is so important to elaborate solar radiation data based on high performance and simple models. In this direction, the main aim of this chapter is to explore the potential of using simple empirical and ANN models to estimate global solar radiation on a horizontal surface. In this regard, Algeria was taken as a case study with eight different locations (Algiers, Oran, Batna, Constantine, Ghardaia, Bechar, Adrar, and Tamanrasset), and four statistical parameters were chosen to assess the performances of each model or approach. Furthermore, the results of these empirical correlations alongside with those of ANN model will be compared, in order to choose the best approach to generate global solar radiation databases in Algeria.

In: Solar Irradiance
Editor: Daryl M. Welsh

ISBN: 978-1-53618-786-1
© 2020 Nova Science Publishers, Inc.

Chapter 1

PHOTOVOLTAIC APPLICATIONS ON THE BASIS OF MODIFIED FUZZY NEURAL NET

Ekaterina A. Engel[*]
Department of Information Technologies and Systems,
Katanov State University of Khakassia, Abakan, Russian Federation

ABSTRACT

The long-term contribution of solar energy is predicated on overcoming remaining technical barriers, mainly through research and development based on artificial intelligence. Data mining and machine learning methods gained a great attention in photovoltaic applications. The real-life photovoltaic systems have complex non-linear dynamic due to random variation of the system parameters and fluctuation of the solar irradiance. Thus, neural-network-based solutions have been proposed to approximate this complex dynamic. But the neuronet needs to become more adaptive. Modification of the network into a recurrent neuronet with fuzzy units provides intelligent behavior. This forms the motivation for the development of intelligent models for sizing, parameters forecasting and control of a photovoltaic system on the basis of a modified fuzzy neural net. Compared to existing fuzzy neuronets, including Adaptive Network-Based Fuzzy Inference System, the modified fuzzy neural net

---

[*] Corresponding Author E-mail: ekaterina.en@gmail.com.

includes recurrent neuronets and fuzzy units. We generated the optimum architecture of the modified fuzzy neural net on the basis of development modified swarm intelligence algorithms. The function approximation capabilities of a neural net are exploited to approximate a membership function. This chapter presents the intelligent models for sizing, parameters forecasting and control of a photovoltaic system on the basis of a modified fuzzy neural net. The modified fuzzy neural net provides automatic fulfillment and modification of all proposed intelligent models.

The first proposed model on the basis of a modified fuzzy neural net provides a two days ahead forecasting of the hourly power from a photovoltaic system under random perturbations. In order to train the effective modified fuzzy neural net we developed the modified multi-dimensional Particle Swarm Optimization, in which the multi-dimensional Particle Swarm Optimization is combined with the Levenberg-Marquardt algorithm. The comparison simulation results show that proposed modified multi-dimensional PSO algorithm outperforms multi-dimensional Particle Swarm Optimization and Levenberg-Marquardt algorithm in training the effective modified fuzzy neural net for a two days ahead forecasting of the hourly PV array power.

The second proposed model on the basis of a modified fuzzy neural net is ambient temperature forecasting model under random perturbations. This model is important for photovoltaic systems. Simulation comparison results for a forecasting of ambient temperature demonstrate the effectiveness of the modified fuzzy neural net trained by proposed modified Ant Lion Optimizer as compared with the same one trained by Ant Lion Optimizer or Levenberg-Marquardt algorithm.

The third proposed model on the basis of a modified fuzzy neural net is a photovoltaic system control model. According the photovoltaic system condition, the modified fuzzy neural net provides a maximum power point tracking under random perturbations. The simulation results demonstrate that the proposed photovoltaic system control model on the basis of a modified fuzzy neural net provides best performance and real-time control speed, as compared to a classical control model with a PID controller based on perturbation & observation, or incremental conductance algorithm.

The fourth proposed model on the basis of a modified fuzzy neural net provides the best configuration and optimal sizing coefficient of a photovoltaic system. The simulation results demonstrate that the effectiveness of the proposed model is better than the Particle Swarm Optimization in sizing of a photovoltaic system.

The fifth intelligent model on the basis of proposed modified Particle Swarm Optimization provides a maximum photovoltaic array power point tracking under fast-changing non-uniform solar irradiance level. Simulation comparison results demonstrate the competitive effectiveness of the proposed modified Particle Swarm Optimization as compared with the perturbation & observation algorithm and Particle Swarm Optimization in finding the global maximum power point of a partially shaded photovoltaic array.

Keywords: photovoltaic system, the maximum power point tracking, the photovoltaic system's control, modified fuzzy neural net, the hourly photovoltaic system power forecasting

ABBREVIATIONS

PV	Photovoltaic
ANFIS	Adaptive Network-Based Fuzzy Inference System
MPP	the maximum power point
MPPT	the maximum power point tracking
GMPP	global maximum power point
V	voltage
DC	direct current
AC	alternating current
AC/DC	alternating current/direct current
PID	Proportional-Integral- Derivative
MoFNN	the Modified Fuzzy Neural Net
MoFNN 1	the Modified Fuzzy Neural Net was trained based on multi-dimensional Particle Swarm Optimization
MoFNN 2	Modified Fuzzy Neural Net was trained based on Levenberg-Marquardt algorithm
MoFNN 3	Modified Fuzzy Neural Net was trained based on modified multi-dimensional Particle Swarm Optimization
RMSE	the root mean square error
W/m^2	Watt per square metre
Ah	Ampere-hour
\$/W	cost per watt
ALO	Ant Lion Optimizer
PSO	Particle Swarm Optimization
Hz	Hertz
kVAR	kilo-volt-ampere reactive

NOMENCLATURE

$G0_i^t$	Extraterrestrial irradiance
Gd_i^t	Historical data of solar irradiance difference
G_i^t	Target solar irradiance
C_{i-m}^{t-2}	Historical data of clear-sky index
Cl_i^t	Cloudiness (%)
PR_i^t	Pressure
W_i^t	Wind speed
Wd_i^t	Wind direction
M	Embedding dimension
$I(V)$	Current-Voltage characteristic
$P(V)$	Power-Voltage characteristic
I	Current
$X(t)$	Voltage
F	Fitness function
$xy_{X,j}^{xd_X(t)}(t)$	j^{th} component (dimension) of the personal best position of ant lion X, in dimension $xd_X(t)$
$xx_{X,j}^{xd_X(t)}(t)$	j^{th} component (dimension) of the position of ant lion X (represents the $j-th$ MoFNN architecture's parameters), in dimension $xd_X(t)$
$vx_{X,j}^{xd_X(t)}(t)$	j^{th} component (dimension) of the position of ant X (represents the $j-th$ MoFNN architecture's parameters), in dimension $xd_X(t)$
$gbest\ (d)$	Global Best position of the elite ant lion index in dimension d
$\hat{xy}_j^d(t)$	j^{th} component (dimension) of the global best position of the elite ant lion, in dimension d
$xd_X(t)$	current dimension of ant lion X position
$vd_X(t)$	dimensional of ant position X
$\tilde{xd}_X(t)$	personal best dimension of ant lion position X

$best(xd_X(t))$	best ant lion
Cumsum	Cumulative sum
Ni	Maximum number of iterations
st	Step of the random walk
r(st)	Stochastic function
M_{Ant}	Position of the ants
A_i	Position of i-th ant
na	Number of ants
$M_{Antlion}$	Vector for each ant lion position
AL_i	Value of the voltage of the i-th ant lion
q	Minimum of the random walk of variable
b	Maximum of the random walk of variable
c^z	Minimum of the variable at the z-th iteration
d^z	Maximum of the variable at the z-th iteration
c^z	Minimum of variable at the z-th iteration
d^z	Maximum of variable at the z-th iteration
$Antlion_e^z$	Position of the selected e-th ant lion at the z-th iteration
z	Current iteration
IT	Maximum number of iterations
w	Constant that depends on the current iteration
R_A^z	Random walk around the ant lion selected by the roulette wheel at the z-th iteration
R_E^z	Random walk around the elite ant lion at the z-th iteration
V	Voltage
ΔV	Voltage increment
dI	Current error before the increment
dV	Voltage error after the increment
Ir	Solar irradiance
P	PV the system's power

S	Input signal of modified fuzzy neural net
m	Air mass
β	Altitude angle
φ_S	Solar azimuth angle
φ_C	Photovoltaic module azimuth angle
p	Reflection factor
Σ	Photovoltaic module tilt angle
C	Sky diffuse factor
A	Parameter related to the Julian day number
k	Parameter related to the Julian day number
G_C	Total rate of radiation striking a photovoltaic system on a clear day
N_{pv}	Total number of photovoltaic modules
N_{pvs}	Number of photovoltaic modules in series
η_{rg}	Efficiency of the regulator (%)
U	Nominal system operating voltage (V)
$[b, m, G_C, I_{SC}, I_0, T_c]$	Dependent variables.
$[N_S, N_P, d, S, f_C]$	Associated control variables
$P_{pv}^{i\,(\max)}(t, S_{opt})$	Maximum power of the photovoltaic module at optimal tilt angle
S_{opt}	Optimal tilt angle
T	Hour, $1 \leq t \leq 24$
i	Day number
$y(T_c)$, $m(V^M, T_c, I^M)$, $\P(T_c)$, $e(m, f_S, b, S, f_C)$, $g(V^M)$: Nonlinear functions	
x_{inv}	Efficiency of the inverter
$P_{load}^i(t)$	Power exhausted by the load at hour t of day i
$C^i(t)$	Available battery capacity (Ah) of day i at hour t
ξ_{bat}	Battery round-trip efficiency (x_{bat} =100% during discharging and x_{bat} = 80% during charging)
V_{Bus}	DC bus voltage
$P_B^i(t)$	Battery input/output power

Dt	Simulation time step
C_{max}	Maximum acceptable storage capacity
C_{min}	Minimum acceptable storage capacity
C_n	Nominal battery's capacity
V_{bus}	Nominal DC bus voltage
V_b	Nominal voltage of each individual battery
P_m	Photovoltaic module's maximum power under standard test conditions
P_{mc}	Battery charger's coefficient
$C_c(u)$	Total capital cost function
$C_m(u)$	Maintenance cost function
u_1	Total number of photovoltaic modules
u_2	Total number of batteries
u	Vector of the cost independent variables $u=(u_1, u_2)$
$L.T_{PV}$	Year life time for photovoltaic module
$L.T_{BAT}$	Year life time for battery
$L.T_{CH}$	Year life time for battery charger
$L.T_{INV}$	Year life time for the inverter
C_{PV}	Capital cost of one photovoltaic module
C_{BAT}	Capital cost of one battery
M_{PV}	Maintenance cost per year of one photovoltaic module
M_{BAT}	Maintenance cost per year of one battery
C_{ch}	Capital cost of one battery charger
y_{ch}	Estimated numbers of the battery charger during the 20 year system lifetime
y_{inv}	The estimated numbers of the DC/AC inverter changings during the 20 year system lifetime
C_{inv}	Capital cost of the inverter
y_{BAT}	Estimated number of battery changings during the 20-year system maintenance
M_{ch}	Maintenance cost per year of one battery charger
M_{inv}	Maintenance cost per year of one DC/AC inverter
N_{PV}	Total optimal number of photovoltaic modules

N_{BAT}	Total optimal number of batteries
O	Algorithm's type
G	Data of solar irradiance
T	Ambient temperature
CP_i	Distribution of the consumer power requirements during a day i
a	Type of photovoltaic modules
$PVmax$	Maximum power from a photovoltaic module
J	Life cycle cost
x	Input signal of MoFNN
N	diode ideality constant
d_{h-1}	the historical data of average monthly ambient temperature
H	number of evaluated samples
K	the Boltzmann constant
Q	the electron charge
a_{Isc}	the temperature coefficient of the short-circuit current
I_L	the light-generated current
I_{Lref}	the photoelectric current under standard condition
$T_{c\,ref}$	module temperature under standard condition
I_{or}	the saturation current
I_o	the reverse saturation current
E_g	the band gap for silicon
D	control signal
g	diode quality factor
R_{sh}	shunt resistance
T_c	effective temperature of the cells
$p_{X,j}$	the personal best position of particle X
$x_{X,j}$	the position of particle X
$v_{X,j}$	the velocity of particle X
e_j	the evaporation rate
g	Global Best position of swarm
w	inertia

c_2	cognitive weight
c_3	social weight
E	the base value of the evaporation rate
G_s	the surface insolation
C_h	the historical data of clear-sky index
$I_i(x)$	the forecasted power of the PV array
$P_i(x)$	the cumulative power of the PV array
$\mu_j(s)$	membership function
$Y(s^i)$	two-layer network
$u(best(d_h))$	best solution X created by the modified multi-dimensional ALO

1. INTRODUCTION

The long-term contribution of solar energy is predicated on overcoming remaining technical barriers, mainly through research and development photovoltaic applications based on artificial intelligence. Data mining and machine learning methods gained a great attention in photovoltaic applications [1-20]. In 2050 PV energy is expected to provide 11% of the global electricity consumption [6].

The real-life photovoltaic systems have complex non-linear dynamic due to random variation of the system parameters and fluctuation of the solar irradiance. Thus, neural-network-based solutions have been proposed to approximate this complex dynamic. But the neuronet needs to become more adaptive. Modification of the network into a recurrent neuronet with fuzzy units provides adaptive behavior. This forms the motivation for the development of intelligent models for sizing, parameters forecasting and control of a photovoltaic system on the basis of a modified fuzzy neural net [8]. Compared to existing fuzzy neuronets, including ANFIS [21], the modified fuzzy neural net includes recurrent neuronets and fuzzy units. The function approximation capabilities of a recurrent neural net are exploited to approximate a membership function. The algorithm of the agent's interaction uses a fuzzy-possibilistic method [7]. It is important to emphasize that a modified fuzzy neural net is improvement of a Multi-Agent Adaptive Fuzzy Neuronet [16]. We generated the optimum architecture of the modified fuzzy neural net based on the proposed algorithm which combines swarm and gradient optimization algorithms.

This chapter presents the intelligent models for sizing, parameters forecasting and control of a photovoltaic system on the basis of a modified fuzzy neural net. The modified fuzzy neural net provides automatic fulfillment and modification of all proposed intelligent models.

The first proposed model on the basis of a modified fuzzy neural net provides a two days ahead forecasting of the hourly power from a photovoltaic system under random perturbations. The PV system power forecasting is critically essential for planning effective transactions in power system operation, because it provides a safety of grid control. The day-ahead market imposes penalties for a deviation from the nominated schedule of the hourly solar plant power. The authors of more recent studies developed the intelligent algorithms which provide PV system power forecasting with good performance [15]. However, there is a growing demand for an accurate PV power forecasting model based on the use of weather forecasting parameters and field measurements. This chapter presents a modified fuzzy neuronet for a two days ahead forecasting of the hourly PV system power. The MoFNN was fulfilled based on an extensive empirical database. A database of the total power from a PV system, ambient temperature, meteorological parameters and insolation data was collected in the south-eastern part of Siberia, Russian Federation at the site of Khakassia. In order to train the effective MoFNN we developed the modified multi-dimensional PSO, in which the multi-dimensional PSO [1] is combined with the Levenberg-Marquardt algorithm [2]. The multi-dimensional PSO provides an area of a global optimum of a modified fuzzy network's architecture, and then for a best solution of iteration we apply the Levenberg-Marquardt algorithm in order to speed up the convergence process. The generation of initial position of a swarm based on Nguyen-Widrow method [23] provides an area of an optimum of a network's architecture at initial iteration and speed up the convergence process. The comparison simulation results leads us to the conclusion that proposed modified multi-dimensional PSO algorithm outperforms multi-dimensional PSO and Levenberg-Marquardt algorithm in training the effective MoFNN for the PV system power forecasting.

The second proposed model on the basis of a modified fuzzy neural net is ambient temperature forecasting model under random perturbations. This model is important for photovoltaic applications. An automatic definition of the optimal architecture's parameters of a neuronet is very complex task which requires an extensive analysis. This forms the motivation to modify evolutionary optimization algorithms such as the Ant Lion Optimizer [24] for detection of an optimum MoFNN architecture. We generated the MoFNN architecture's parameters (an agent's number, a number of nodes in hidden layer, corresponded weights and biases) from the global optimum. In order to generate the optimum modified fuzzy neuronet we developed modified multi-dimensional ALO algorithm, in which we modified the ALO and combined it with the Levenberg-Marquardt algorithm. The multi-dimensional modification of an ALO provides an area of a global optimum of a modified fuzzy network's architecture, and then for a best solution of iteration we apply the Levenberg-Marquardt algorithm in order to speed up the convergence process. The generation of initial personal best and best positions of a swarm based on Nguyen-Widrow method [23] provides an area of an optimum of a network's architecture at initial iteration and speed up the convergence process. The simulation results demonstrate that the proposed modified multi-dimensional ALO outperforms ALO and Levenberg-Marquardt algorithm in training the optimum MoFNN for average monthly ambient temperature forecasting.

The third proposed model on the basis of a modified fuzzy neural net is a photovoltaic system control model. PV system is non-linear and commonly suffers from restrictions imposed by sudden variations in the solar irradiance level. Within the research literature, a whole array of differing MPPT algorithms has been proposed [13]. Among them, the perturbation & observation and incremental conductance algorithms are the most common due to simplicity and easy implementation. But controllers based on P&O, or IC algorithm for PV systems have slow response times to changing reference commands, take considerable time to settle down from oscillating around the target reference state, must often be designed by hand. Therefore, automatic intelligent algorithms such as fuzzy neural networks are promising alternatives [14]. This forms the motivation for the development of a PV system control model on the basis of a modified fuzzy neural net. According the photovoltaic system condition, the modified fuzzy neural net provides a maximum power point tracking under random perturbations. The simulation results demonstrate that the proposed photovoltaic system control model on the basis of a modified fuzzy neural net provides best performance and real-time control speed, as compared to a classical control model with a PID controller based on perturbation & observation, or incremental conductance algorithm.

The fourth proposed model on the basis of a modified fuzzy neural net provides the best configuration and optimal sizing coefficient of a photovoltaic system. Within the research literature, a whole array of differing sizing methods for a photovoltaic system has been proposed [2, 3]. Optimal sizing of photovoltaic systems is a very complex task which requires the development of mathematical models for the photovoltaic system's components as well as usage of global optimization methods. This research solves the task of a photovoltaic system's sizing on the basis of a modified fuzzy neural net. The experimental results demonstrate that the modified fuzzy neural net provides a better solution than does a PSO. The comparison results demonstrate that the modified fuzzy neural net provided optimum solar and battery ratings.

The fifth intelligent model on the basis of proposed modified PSO provides a maximum PV array power point tracking under fast-changing non-uniform solar irradiance level. In such a condition several maximum power points exist but we need to find only one global maximum. The proposed modified PSO employs re-randomization and the particles around the maximum power point initialization. This efficient initialization of particles provides avoiding both unnecessary and redundant searching and a situation in which the area of swarm search doesn't include a global maximum. As a result, this significantly reduces the time that wasted by searching GMPP in the wrong area. Simulation comparison results demonstrate that proposed modified PSO significantly reduces the computational time, as compared to a PSO. This is a huge improvement upon the conventional PSO algorithm which provides the new operating point too far from the MPP and requires more iteration to reach GMPP. Simulation comparison results demonstrate the competitive effectiveness, a maximum PV system power and real-time control speed of the proposed modified PSO as compared to a perturbation & observation algorithm and PSO in finding the global maximum power point of a partially shaded photovoltaic array.

2. PHOTOVOLTAIC ARRAY POWER FORECASTING SYSTEM MODEL ON THE BASIS OF A MODIFIED FUZZY NEURAL NET

Researches of a photovoltaic system which integrated in electric power systems gained a great attention in modern energetics. The PV system power forecasting is critically essential for planning effective transactions in power system operation, because it provides a safety of grid control. The day-ahead market imposes penalties for a deviation from the nominated schedule of the hourly solar plant power. The authors of more recent studies developed the intelligent algorithms which provide PV system power forecasting with good performance. However, there is a growing demand for an accurate PV power forecasting model based on the use of weather forecasting parameters and field measurements. This chapter presents a modified fuzzy neuronet for a two days ahead forecasting of the hourly PV system power. The MoFNN was fulfilled based on an extensive empirical database. A database of the total power from a PV system, ambient temperature, meteorological parameters and insolation data was collected at the site of Khakassia in the south-eastern part of Siberia, Russian Federation. In order to train the effective MoFNN we developed the modified multi-dimensional PSO, in which the multi-dimensional PSO [1] is combined with the Levenberg-Marquardt algorithm [2]. The multi-dimensional PSO provides an area of a global optimum of a modified fuzzy network's architecture, and then for a best solution of iteration we apply the Levenberg-Marquardt algorithm to speed up the convergence process. The generation of initial positions of a swarm based on Nguyen-Widrow method [23] provides an area of an optimum of a network's architecture at initial iteration and speed up the convergence process. The comparison simulation results leads us to the conclusion that proposed modified multi-dimensional PSO algorithm outperforms multi-dimensional PSO and Levenberg-Marquardt algorithm in training the effective MoFNN for the PV system power forecasting.

2.1. The Power from a PV Array

In this research we investigate a 5 KW Solar Power Plant (model SA-5000M). This PV array places at the site of Khakassia (91.4° of longitude East, 53.7° of latitude North and 246 m of altitude). Figure 1 demonstrates a scheme of the PV array. This PV array includes six solar PV modules (CHN250-60P), a solar regulator, a battery bank and an inverter (Figure 1).

The total rate of radiation G_C striking a PV array on a clear day calculates as follows:

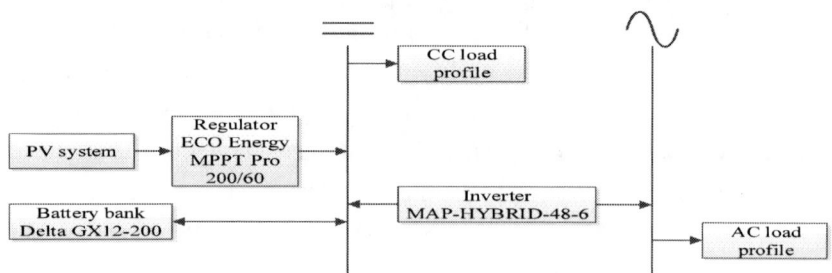

Figure 1. The scheme of the PV array.

$$Gc = Ae^{-km}(\cos\beta\cos(\phi_s - \phi_c)\sin\Sigma + \sin\beta\cos\Sigma + (C + \cos\Sigma)/2 + p(\sin\beta + C)(1 - \cos\Sigma)/2) \quad (1)$$

where m is the air mass, β is the altitude angle, φ_S is the solar azimuth angle, φ_C is the PV module azimuth angle, p is the reflection factor, Σ is the PV module tilt angle, C is the sky diffuse factor, A and k are parameters related to the Julian day number.

The surface irradiance is less than its corresponding extraterrestrial irradiance. The cloudiness defines a degree of attenuation. The surface irradiance fluctuates randomly. The cloudiness' dynamic defines these fluctuations (Figure 2).

We use a clear-sky index due to evaluate influences of the deterministic solar geometry and the nondeterministic atmospheric extinction separately. In this chapter, we define the clear-sky index as follows:

$$C = Gs / Gc, \quad (2)$$

where Gs is the surface insolation, Gc is the clear-sky model's insolation. The insolation is the integral of solar irradiance over a time period.

Figure 3 demonstrates that the clear-sky index C is big and has similar shape on sunny days (05/18/16, 05/19/16) at the site of Khakassia. In contrast, C is smaller and has more fluctuations on cloudy days (05/16/16, 05/17/16).

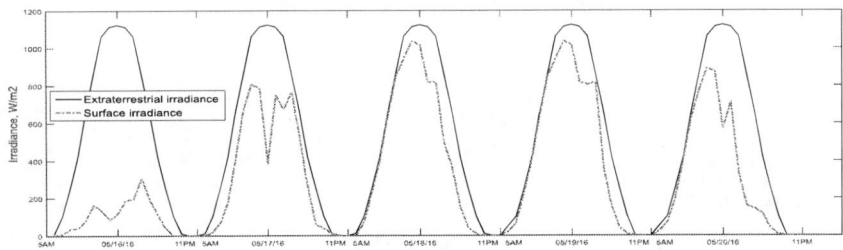

Figure 2. The extraterrestrial irradiance and the surface irradiance at the site of Khakassia.

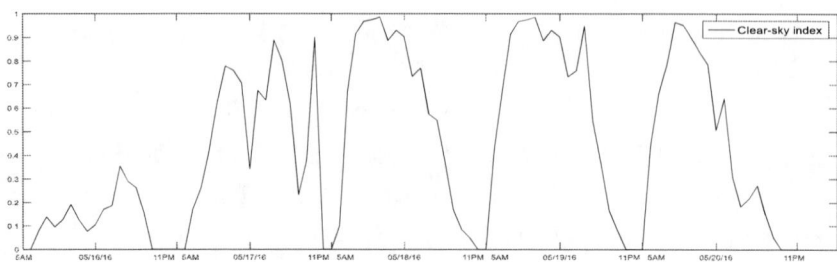

Figure 3. The clear-sky index at the site of Khakassia.

We fulfilled the MoFNN $Fes_{jhq}(x_h^t)$ based on the data

$$s_h^t = (x_h^t = (G0_h^t, C_h^{t-2}, P_h^{t-2}, Cl_h^t, T_h^t, PR_h^t, W_h^t, Wd_h^t), P_h^t), \qquad (3)$$

where $G0_h^t$ is the extraterrestrial irradiance, P_h^t is the power from a PV array, P_h^{t-2} is the historical data of the power from a PV array, C_h^{t-2} is the historical data of clear-sky index, Cl_h^t is the cloudiness (%), PR_h^t is the pressure, W_h^t and Wd_h^t are the wind speed and the wind direction, respectively, T_h^t is the ambient temperature, $t= \overline{5..23}$, $t=\overline{1..1217}$. It is to emphasize that Cl_h^t, PR_h^t, W_h^t, Wd_h^t, T_h^t are daily average parameters of the weather forecast. The number of samples is 23123 ($h*t = 19*1217 = 23123$). This database was collected at the site of Khakassia from March 2016 through July 2019.

2.2. The Training of the Modified Fuzzy Neural Net

The main purpose of this research is to create the effective MoFNN for hourly PV array power forecasting. In order to train the effective modified fuzzy neural net for hourly PV power forecasting we developed the modified multi-dimensional PSO (Figure 4) which combined multi-dimensional PSO and the Levenberg-Marquardt algorithm. The multi-dimensional PSO globally optimizes the network's structure, and then for a best solution of iteration we apply the Levenberg-Marquardt algorithm in order to speed up the convergence process. The generation of initial positions of a swarm based on Nguyen-Widrow method [23] provides an area of an optimum of a network's architecture at initial iteration. We fulfilled the agents of the MoFNN as two-layered recurrent networks. We coded the MoFNN architecture's parameters (weights and biases) into particles a. The dimension range of the proposed modified multi-dimensional PSO and multi-dimensional PSO is $[D_{\min}, D_{\max}]$, $D_{\min} = 34, D_{\max} = 144$ (Figure 4).

MD PSO (termination criteria: $\{IterNo, \varepsilon_C, \ldots\}$)

Step 1. For $\forall a \in \{1, S\}$ do:

Randomize $xd_a(0)$: Dimension component of particle a, $vd_a(0)$: Velocity component of dimension of particle a,

Initialize $x\tilde{d}_a(0) = xd_a(0)$

$x\tilde{d}_a$: Personal best dimension component of particle a, For $\forall d \in \{D_{\min}, D_{\max}\}$ do:

We generates $xx_a^d(0)$ and $xv_a^d(0)$ based on Nguyen-Widrow method [23], Initialize $xy_a^d(0) = xx_a^d(0)$. End For. End For.

Step 2. For $\forall t \in \{1, IterNo\}$ do:

For $\forall a \in \{1, S\}$ do:

If $(f(xx_a^{xd_a(t)}(t)) < \min(f(xy_a^{xd_a(t)}(t-1)), \min_{p \in S - \{a\}}(f(xx_p^{xd_a(t)}(t)))))$ then do $xy_a^{xd_a(t)}(t) = xx_a^{xd_a(t)}(t)$

If $f(xx_a^{xd_a(t)}(t)) < f(xy_{gbest(xd_a(t))}^{xd_a(t)}(t-1))$ then $gbest(xd_a(t)) = a$, $gbest(d)$: Global best particle index in dimension d,

If $f(xx_a^{xd_a(t)}(t)) < f(xy_a^{x\tilde{d}_a(t-1)}(t-1))$ then $xd_a(t) = xd_a(t)$

If $f(xx_a^{xd_a(t)}(t)) < f(x\hat{y}_{dbest}(t-1))$ then $dbest = xd_a(t)$ End If. End For.

If the termination criteria are met, then Stop. $xy_{a,j}^{xd_a(t)}(t-1)$: j[th] component of the personal best position of particle a, in dimension $xd_a(t)$

$\mu = vd_a(t)$.

Step 3. $E = f(xx_{gbest(xd_a(t))}^{xd_a(t)})$

Step 4. While $(I < IterNo) OR (E > \varepsilon_c)$ $\Delta W_I = [J_I^T J_I + \mu * X]^{-1} J_I^T E$, where J_I is the Jacobian matrix, μ is the learning rate

Step 5. Calculate $W_I = W_I + \Delta W_I$, $xx_{gbest(xd_a(t))}^{\prime xd_a(t)}$, $E' = f(xx_{gbest(xd_a(t))}^{\prime xd_a(t)})$

Step 6. If $E' < E$ then $W_I = W_I + \Delta W_I$; $\mu = \mu\beta$, $E' = E$; Go to step 3 else $\mu = \mu / \beta$; go to step 5 endIf.

Step 7. For $\forall a \in \{1, S\}$ do:

For $\forall j \in \{1, xd_a(t)\}$ do:

Compute $vx_{a,j}^{xd_a(t)}(t+1) = w(t) vx_{a,j}^{xd_a(t)}(t) + c_1 r_{1,j}(t)(xy_{a,j}^{xd_a(t)}(t) - xx_{a,j}^{xd_a(t)}(t)) + c_2 r_{2,j}(t)(x\hat{y}_j^{xd_a(t)}(t) - xx_{a,j}^{xd_a(t)}(t))$,

$$xx_{a,j}^{xd_a(t)}(t+1) = \begin{cases} xx_{a,j}^{xd_a(t)}(t) + vx_{a,j}^{xd_a(t)}(t+1) & if\ X_{\min} \le vx_{a,j}^{xd_a(t)}(t+1) \le X_{\max} \\ U(X_{\min}, X_{\max}) + xx_{a,j}^{xd_a(t)}(t+1) & else \end{cases},$$

$$xx_{a,j}^{xd_a(t)}(t+1) \leftarrow \begin{cases} xx_{a,j}^{xd_a(t)}(t+1) & if\ X_{\min} \le xx_{a,j}^{xd_a(t)}(t+1) \le X_{\max} \\ U(X_{\min}, X_{\max}) & else \end{cases}$$

. Update velocities and locations in other dimensions. End For.

Compute $vd_a(t+1) = \lfloor vd_a(t) + c_1 r_1(t)(x\tilde{d}_a(t) - xd_a(t)) + c_2 r_2(t)(dbest - xd_a(t)) \rfloor$,

$$xd_a(t+1) = \begin{cases} xd_a(t) + vd_a(t+1) & if\ VD_{\min} \le vd_a(t+1) \le VD_{\max} \\ xd_a(t) + VD_{\min} & if\ vd_a(t+1) < VD_{\min} \\ xd_a(t) + VD_{\max} & if\ vd_a(t+1) > VD_{\max} \end{cases}, \quad xd_a(t+1) \leftarrow \begin{cases} xd_a(t) & if\ P_d(t+1) \ge \max(15, xd_a(t+1)) \\ xd_a(t) & if\ xd_a(t+1) < D_{\min} \\ xd_a(t) & if\ xd_a(t+1) > D_{\max} \\ xd_a(t+1) & else \end{cases}$$

. End For. End For

Figure 4. A modified multi-dimensional Particle Swarm Optimization.

for each $agent_q$ in subculture S_k do $g_{khq}(x_h^t) \leftarrow$ GetResponse($agent_q; x_h^t$);
$v_q \leftarrow$ TakeAction($g_{khq}(x_h^t)$): Evaluate e_q as (4); $v_q = 1 - e_q$.
end for. $w=[g_{kh1}(x_h^t), ..., g_{khq}(x_h^t)]$ Calculate $I_h = Fes_{jh}(g_{jhq}(s))$ based on ($w, [v_1, ..., v_q]$) as fuzzy expected solution (Fes) in 2 steps [3]

Step 1: Solve equation $\left[\prod_{i=1}^{q}(1+\lambda w_i)-1\right]/\lambda = 1, \quad -1 < \lambda < \infty.$

Step 2: Calculate $s = \int h \circ W_\lambda = \sup_{\alpha \in [0,1]} \min\{\alpha, W_\lambda(F_\alpha(v_j))\}$, where $F_\alpha(v_j) = \{F_i | F_i, v_j \geq \alpha\}, v_j \in V,$

$W_\lambda(F_\alpha(v_j)) = \left[\prod_{F_i \in F_\alpha(v_j)}^{k}(1+\lambda w_i)-1\right]/\lambda.$ Calculate $I_h = \max_{v_i \in V} s(w_j)$

Figure 5. Algorithm of the agent's interaction.

I unit: Training of the MoFNN briefly can be described as follows
All samples ($N= h*t=19*1217=23123$) were classified into two groups: A_1 – sunny hour ($T_h^t=1$), A_2 – cloudy ($T_h^t=-1$). This classification generates vector with elements T_h^t. Two-layer recurrent network (number of hidden neurons and delays are 7 and 2, respectively): $F(X)$ was trained. The vector $X=x_h^t$ was network's input. The vector T_h^t was network's target. Fuzzy sets A_j (A_1 – sunny hour, A_2 – cloudy) with membership function $\mu_j(x)$ are formed base on aforementioned two-layer recurrent network $F(x_h^t)$, $j = \overline{1.2}$.
We train based on an optimization algorithm o (if $o=1$ then optimization algorithm is multi-dimensional PSO, if $o=2$ then optimization algorithm is Levenberg-Marquardt algorithm, if $o=3$ then optimization algorithm is the proposed algorithm) three two-layered recurrent neural networks: $g_{jhq}(x_h^t)$, $h = \overline{1.19}$, $j = \overline{1.2}$, $q = \overline{1.3}$, based on the data (3). This step provides recurrent neural networks which create the forecasted power of the PV system $g_{jhq}(x_h^t)$. Two agent's subcultures S_j are formed base on aforementioned two-layer recurrent networks.
If-then rules are defined as: Π_j: IF X is A_j THEN $I_h = Fes_{jh}(g_{jhq}(x_h^t))$. (5)
II unit: Simulation of the trained MoFNN $\forall c \in \{23092..23123\}$ for $h=1:19$
Aggregation antecedents of the rules (5) maps input data x_h^c into their membership functions and matches data with conditions of rules. These mappings are then activates the k rule, which indicates the k hour's state $k = \overline{1.2}$ and k agent's subcultures – S_k.
According the k hour's state the MoFNN (trained base on the data s_h^d, where $d = \overline{1.c-1}$) creates the forecasted power of the PV system $I_h = Fes(f_{jhq}(x_h^c))$ as a result of multi-agent interaction (Fig. 5) of subculture S_k.

Figure 6. Fulfilment of the MoFNN.

In this research we define a fitness function $f(x)$ based on the Chebyshev criterion as follows

$$e = f(x) = \max_{i=1..N}\left(\left|P_i(x) - I_i(x)\right|/P_i(x)\right), \qquad (4)$$

where N is the number of data samples, $I_i(x)$ is the forecasted power of the PV array, $P_i(x)$ is the cumulative power of the PV array.

The algorithm of the agent's interaction (Figure 5) uses a fuzzy-possibilistic method [7]. Fulfilment of the MoFNN briefly can be described by Figure 6.

2.3. Simulation Results of a Two Days Ahead PV Power Hourly Forecasting Model on the Basis of a Modified Fuzzy Neural Net

To illustrate the benefits of the MoFNN in two days ahead forecasting of the hourly power from the PV array, the numerical examples from the previous subsections 2.1 and 2.2 are revisited using the software [11]. The three MoFNN based on the training set of the data (3) s_h^d were fulfilled ($d=\overline{1..c-1}$, $\forall c \in \{23092..23123\}$). The first MoFNN1 was trained using multi-dimensional PSO ($o=1$). We trained the second MoFNN (MoFNN2, $o=2$) based on Levenberg-Marquardt algorithm. We trained the third MoFNN (MoFNN3, $o=3$) based on the modified MD PSO. Due to obtain statistical results, we perform 120 modified MD PSO and modified MD PSO runs with following parameters: $S=250$ (we use 250 particles), $E = 150$ (we terminate at the end of 150 epochs). We evaluated the forecast accuracies of the aforementioned models as the fitness function (4). Table 1 demonstrates that only one set of MoFNN3 architecture with *dbest* = 56 can achieve the fitness function (4) under *4,8%* over the holdout set of the data (3), $\forall c \in \{23092..23123\}$.

We chose MoFNN3 solution with *dbest=56* as an optimum modified fuzzy neuronet. The MoFNN3 has three agents of each subculture S_k. The aforementioned agents are the two-layered recurrent neural network. The first and second agent's number of hidden neurons and delays are 2. The third agent's number of hidden neurons and delays are 1 and 2, respectively. MoFNN2 has same architecture. Figure 7 demonstrates the mean convergence curves of the multi-dimensional PSO and the proposed algorithm for training a MoFNN. The generation of initial positions of a swarm based on Nguyen-Widrow method [23] provides an area of an optimum of a network's architecture at initial iteration and speed up the convergence process.

Figure 7 demonstrates that the MoFNN3 has definitely more convergence speed over training set of the data s_h^d (3) ($d = \overline{1..c-1}$, $\forall c \in \{23092..23123\}$) than the MoFNN1 in the PV array power forecasting. The generation of initial positions of a swarm based on Nguyen-Widrow method [23] provides an area of an optimum of a network's architecture at initial iteration and speed up the convergence process.

Table 2 demonstrates that errors (4) of the three MoFNNs in sunny hours are quite small.

Table 1. results of modified multi-dimensional PSO

The MoFNN's dbest dimension	36	46	56	66	76	86	106	116	126	136	146
The fitness function (4) (%)	4,93	4,90	4,78	4,92	4,93	4,95	4,99	5,00	5,02	5,06	5,07

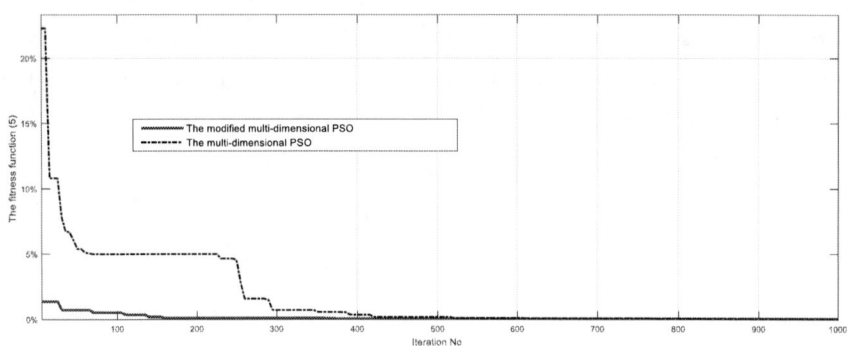

Figure 7. The mean convergence curves.

The performances of the MoFNN1 and the MoFNN3 are changing in sunny and cloudy hours (Table 2).

Table 2. A two days ahead forecasting of the hourly power from the PV array: comparison of results

The MoFNN with dbest = 56	MoFNN3 solution		MoFNN1 solution		MoFNN2 solution	
	Sunny	Cloudy	Sunny	Cloudy	Sunny	Cloudy
The fitness function (4) (%)	3,81	4,71	3,84	4,78	4,71	5,88

Nevertheless, the MoFNN1 and the MoFNN3 effectively track the complex dynamics of real measured data in cloudy hours. Table 2 indicates that the MoFNN3 outperform the MoFNN2 and the MoFNN1, especially in the cloudy hours. The performance of the MoFNN3 trained by proposed algorithm is superior to the same one trained by multi-dimensional PSO or Levenberg-Marquardt algorithm, especially during fast variations of cloudiness. Simulation comparison results for a two days ahead forecasting of the hourly PV array power demonstrate the effectiveness of the MoFNN trained by the proposed algorithm as compared with the same ones trained by multi-dimensional PSO or Levenberg-Marquardt algorithm. The analysis of the evolving errors demonstrates the potential of the MoFNN in the hourly PV array power forecasting.

3. TEMPERATURE FORECASTING MODEL ON THE BASIS OF A MODIFIED FUZZY NEURAL NET

The average monthly ambient temperature forecasting is important for different reasons in multiple areas, including PV applications. We present a modified fuzzy neuronet for average monthly ambient temperature forecasting. We fulfilled the agents of the MoFNN based on neural networks. An automatic definition of the optimal architecture's parameters of a neuronet is very complex task which requires an extensive analysis of the system and the trial-error process. This process is demanding because it is difficult to anticipate all conditions of optimal neuronet architecture. This forms the motivation to modify evolutionary optimization techniques such as the Ant Lion Optimizer [24] for detection of an optimum MoFNN architecture. We generated the MoFNN architecture's parameters (an agent's number, a number of nodes in hidden layer, corresponded weights and biases) from the global optimum. In order to generate the optimum modified fuzzy neuronet we developed modified multi-dimensional ALO algorithm, in which we modified the ALO and combined it with the Levenberg-Marquardt algorithm. We first modified ALO to globally optimize the modified fuzzy network's structure in multi-dimensional space, and then we elaborated the Levenberg-Marquardt algorithm to speed up the convergence process. The multi-dimensional modification of an ALO provides an area of a global optimum of a modified fuzzy network's architecture, and then for a best solution of iteration we apply the Levenberg-Marquardt algorithm in order to speed up the convergence process. The generation of initial personal best and best positions of a swarm based on Nguyen-Widrow method [23] provides an area of an optimum of a network's architecture at initial iteration and speed up the convergence process. The simulation results demonstrate that proposed modified multi-dimensional ALO algorithm outperforms ALO and Levenberg-Marquardt algorithm in training the optimum MoFNN for average monthly ambient temperature forecasting.

3.1. The Modified Fuzzy Neuronet for Average Monthly Ambient Temperature Forecasting

The modified fuzzy neuronets are fulfilled based on the data

$$Z_h=(\ x_h^0=(d_{h-1},\ d_{h-2},\ d_{h-3},\ d_{h-4},\ d_{h-5},\ d_{h-6}), x_h^1=(p^1{}_h,\ p^2{}_h), x_h^2=(d_{h-1},\ d_{h-6}), d_h), \quad (5)$$

where d_{h-i} is the historical data of average monthly ambient temperature; $p^1{}_h$, $p^2{}_h{}^t$ are the first two principal components of the x_h^0 (the first three principal component variances are 49.3, 48.9 and 0.6, respectively), $h=\overline{1..1626}$, $i=\overline{0..6}$. This database was collected at the site of Irkutsk from 01/1882 through 12/2017.

The agents of the MoFNN $F_{jq}(x_h^z)$ are fulfilled as two-layered networks, $z=\overline{1..2}$, $j=\overline{1..2}$. We coded the MoFNN architecture's parameters (an agent's number – q, number of nodes in hidden layer, corresponded weights and biases) into particles X. For ALO and modified multi-dimensional ALO algorithms these particles X represented the swarm of ants and ant lions. The fitness function evaluated as follows:

$$f(X, x^z) = (1/H) \sum_{l=1}^{H} |d_l - F_{jq}(X, x_l^z)|, \qquad (6)$$

where H is number of evaluated samples. We used function (6) as a fitness function for the modified multi-dimensional ALO and ALO algorithms.

3.2. The Modified Multi-Dimensional ALO Algorithm for Training the MoFNN

We represented the modified multi-dimensional ALO process at $t-th$ iteration by the following characteristics:

$xy_{X,j}^{xd_X(t)}(t)$: j^{th} component (dimension) of the personal best position of ant lion X, in dimension $xd_X(t)$;

$*x_{X,j}^{xd_X(t)}(t)$: j^{th} component (dimension) of the position of ant lion (*=x)/ ant (*=v) X (represents the $j-th$ MoFNN architecture's parameters), in dimension $xd_X(t)$;

gbest (d): Global Best position of the elite ant lion index in dimension d;

$x\hat{y}_j^d(t)$: j^{th} component (dimension) of the global best position of the elite ant lion, in dimension d;

$xd_X(t)$: current dimension of ant lion X position;

$vd_X(t)$: dimensional of ant position X;

$\tilde{xd}_X(t)$: personal best dimension of ant lion position X;

$best(xd_X(t))$: best ant lion.

The function Randomize generates random MoFNN architecture's parameters values (all walks of ants) based on the following equation.

$$vx_{X,j}^{xd_X(t)}(t) = X = [0, cumsum(2r(t_1)-1), \ldots, cumsum(2r(t_T)-1)], \quad (7)$$

where *cumsum* is the cumulative sum, t demonstrates the step of the random walk, T is the maximum number of iterations, $r(t)$ is a stochastic function defined as follows:

$$r(t) = \begin{cases} 1 & if \quad rand > 0.5 \\ 0 & if \quad rand \leq 0.5 \end{cases} \quad (8)$$

In order to preserve random walks of ants inside the search space, they are normalized in the following way:

$$vx_X^{xd_X(t)}(t) = c + \left(vx_X^{xd_X(t)}(t) - a\right) \times \left(b - c'\right) / \left(d' - a\right), \quad (9)$$

where for this research *min(a)* = -1 is the minimum of the random walk of variable, *max(a)* = 1 is the maximum of the random walk of variable, c' is the minimum of the variable at the $t-th$ iteration, and d' is the maximum of the variable at the t-th iteration. Mathematical modeling of ants trapping in the ant lion's pits is given as follows:

$$c_i^t = xx_j^{xd_X(t)}(t) + c^t, \quad d_i^t = xx_j^{xd_X(t)}(t) + d^t, \quad (10)$$

where c' is the minimum of variable at the t-th iteration, d' is the maximum of variable at the t-th iteration, and $vx_j^{xd_X(t)}(t)$ is the position of the selected

j-th ant lion at the *t-th* iteration. The ant lion's hunting capability is modeled by fitness proportional roulette wheel selection. The mathematical model that describes the way the trapped ant slides down towards the ant lion is given as follows:

$$c^t = c^t \cdot T/(10^W \cdot t), \quad d^t = d^t \cdot T/(10^W \cdot t), \tag{11}$$

where t is the current iteration and w is the constant that depends on the current iteration.

We expressed the modified multi-dimensional ALO algorithm for training the MoFNN (the termination criteria is $\{T, \varepsilon_C, ...\}$; S is the number of ants and ant lions) as follows:

1. For $\forall X \in \{1, S\}$ do:
   1.1. Randomize $xd_X(1)$, $vd_X(1)$.
   1.2. Initialize $\widetilde{xd}_X(0) = xd_X(1)$.
   1.3. For $\forall d=4*h+3 \in \{D_{min}, D_{max}\}$ do:
      3.1.1. Randomize $xx_X^d(1)$, $vx_X^d(1)$. The function Randomize generated the position of the ants $vx_X^d(1)$ by usage (8)-(12).
      1.3.2 We generate of initial personal best position of a swarm $xy^d(1)$ and best position of a swarm $\hat{xy}^d(1)$ based on Nguyen-Widrow method [23].
   1.4. End For.
2. End For.
3.. For $\forall t \in \{1, T\}$ do:
   3.1. For $\forall X \in \{1, S\}$ do:
      3.1.1. If $f(xx_X^{xd_X(t)}(t)) > f(xy_X^{xd_X(t-1)}(t-1))$
         3.1.1.1. then Do: $xy_X^{xd_X(t)}(t) = xx_X^{xd_X(t)}(t)$
         3.1.1.2. If $f(xx_X^{xd_X(t)}(t)) < f(xy_X^{\widetilde{xd}_X(t-1)}(t-1))$
            3.1.1.2.1. then $xd_X(t) = xd_X(t-1)$
         3.1.1.3. End If
         3.1.1.4. else $xd_X(t) = xd_X(t)$
      3.1.2. else $xy_X^{xd_X(t)}(t) = xy_X^{xd_X(t)}(t-1)$
   3.1.3. End If
   3.1.4. If $(f(xx_X^{xd_X(t)}(t)) < \max(f(xy_X^{xd_X(t)}(t-1)), \max_{1 \le p < X}(f(xx_p^{xd_X(t)}(t))))$

3.1.4.1. then Do: $gbest(xd_X(t)) = X$. For the t-th iteration the best ant lion $best(xd_X(t)) = X$ is considered as elite. This rule is fulfilled elitism. The elitism indicates that every ant randomly walks near selected the ant lion and has a position $vx_{X_i}^{xd_X(t)}(t) = (R_A^t + R_E^t)/2$, where R_A^t is the random walk around the ant lion selected by the roulette wheel at the t-th iteration and R_E^t is the random walk around the elite ant lion at the t-th iteration.

3.1.4.2. If $f(xx_X^{xd_X(t)}(t)) > f(x\hat{y}^{dbest}(t-1))$

3.1.4.2.1. then $dbest = xd_X(t)$

3.1.4.3. End If

3.1.5. End If

3.1.6. In other dimensions $\forall d \in [D_{\min}, D_{\max}] - \{xd_X(t)\}$ do updates $xy_{X,j}^d(t) = xy_{X,j}^d(t-1)$, $x\tilde{y}_j^d(t) = x\tilde{y}_j^d(t-1)$.

3.2. End For

3.3. If the termination criteria are met

3.3.1. then Stop.

3.4. End If

3.4.1. $\mu = vd_a(i)$.

3.4.2. $E = f(xx_{gbest(xd_a(i))}^{xd_a(i)})$

3.4.3. While ($I < IterNo$) OR ($E > \varepsilon_c$) $\Delta W_I = [J_I^T J_I + \mu * X]^{-1} J_I^T E$, where J_I is Jacobian matrix, μ is learning rate.

3.4.4. Calculate $W_I = W_I + \Delta W_I$, $xx_{gbest(xd_a(i))}'^{xd_a(i)}$, $E' = f(xx_{gbest(xd_a(i))}'^{xd_a(i)})$.

3.4.5. If $E' < E$ Then $W_I = W_I + \Delta W_I$; $\mu = \mu\beta$; $E' = E$; Go to 3.4.2 Else $\mu = \mu/\beta$; Go to 3.4.4 End If.

3.5. For $\forall X \in \{1, S\}$ do:

3.5.1. For $\forall j \in \{1, xd_X(t)\}$ do:

3.5.1.1. Compute

$$vx_{X,j}^{xd_X(t)}(t+1) = w(t)vx_{X,j}^{xd_X(t)}(t) + c_1 r_{1,j}(t)(xy_{X,j}^{xd_X(t)}(t) - xx_{X,j}^{xd_X(t)}(t)) + c_2 r_{2,j}(t)(xy_j^{\widehat{xd_X(t)}}(t) - xx_{X,j}^{xd_X(t)}(t)),$$

$$xx_{X,j}^{xd_X(t)}(t+1) = \begin{cases} xx_{X,j}^{xd_X(t)}(t) + vx_{X,j}^{xd_X(t)}(t+1) & if\ X_{\min} \leq vx_{X,j}^{xd_X(t)}(t+1) \leq X_{\max} \\ U(X_{\min}, X_{\max}) + xx_{X,j}^{xd_X(t)}(t+1) & else \end{cases},$$

$$xx_{X,j}^{xd_X(t)}(t+1) \leftarrow \begin{cases} xx_{X,j}^{xd_X(t)}(t+1) & if\ X_{\min} \leq xx_{X,j}^{xd_X(t)}(t+1) \leq X_{\max} \\ U(X_{\min}, X_{\max}) & else \end{cases}$$

3.5.1.2. In other dimensions $\forall d \in [D_{\min}, D_{\max}] - \{xd_X(t)\}$ do updates locations of ants and ant lions $vx_{X,j}^d(t+1) = vx_{X,j}^d(t)$,

$$xx_{X,j}^d(t+1) = xx_{X,j}^d(t).$$

3.5.2. End For

3.5.3. Compute

$$vd_X(t+1) = \left| vd_X(t) + c_1 r_1(t)(\widetilde{xd}_X(t) - xd_X(t)) + c_2 r_2(t)(dbest - xd_X(t)) \right|,$$

$$xd_X(t+1) = \begin{cases} xd_X(t) + vd_X(t+1) & if\ VD_{\min} \leq vd_X(t+1) \leq VD_{\max} \\ xd_X(t) + VD_{\min} & if\ vd_X(t+1) < VD_{\min} \\ xd_X(t) + VD_{\max} & if\ vd_X(t+1) > VD_{\max} \end{cases},$$

$$xd_X(t+1) \leftarrow \begin{cases} xd_X(t) & if\ P_d(t+1) \geq \max(15, xd_X(t+1)) \\ xd_X(t) & if\ xd_X(t+1) < D_{\min} \\ xd_X(t) & if\ xd_X(t+1) > D_{\max} \\ xd_X(t+1) & else \end{cases}$$

3.6. End For

4. End For

The modified multi-dimensional ALO generated the optimum MoFNN architecture's parameters (an agent's number, a number of nodes in hidden layer, corresponded weights and biases).

3.3. Fulfillment of the MoFNN

In order to train the optimum agents of the MoFNN for average monthly ambient temperature forecasting we modified ALO to globally optimize the network's structure (the modified ALO will stop after a global solution is localized within small region), and then we used Levenberg-

Marquardt to speed up convergence process. The algorithm of the agent's interaction elaborates a fuzzy-possibilistic method and includes three steps:

Step 0: for each $agent_q$ in $subculture$ S_k do

$F_{kq}(x_h^z) \leftarrow$ GetResponse($agent_q$; x_h^z);

$v_q \leftarrow$ TakeAction($F_{kq}(x_h^z)$):

Evaluate $v_q = 1 - f(w, x_H^z)$, where f is the objective function (6), x_H^z is a sample which we chose from data (5) according condition: $\min_{i \in 1..N} |x_i^z - x_H^z|$.

end for.

$w = [F_{k1}(x_h^z), ..., F_{kq}(x_h^z)]$.

We compute $I_h = Fes_{kh}(F_{kq}(x_h^z))$ based on $(w, [v_1, ..., v_q])$ as fuzzy expected solution (Fes) in 2 steps [7]:

Step 1: Solve equation $\left[\prod_{i=1}^{q}(1+\lambda w_i) - 1\right]/\lambda = 1$, $-1 < \lambda < \infty$.

Step 2: We compute $s = \partial t \circ W_l = \sup_{a \uparrow [0,1]} \min\{a, W_l(F_a(v_j))\}$, where

$F_a(v_j) = \{F_i | F_i, v_j \ni a\}, v_j \hat{I} V$, $W_\lambda(F_\alpha(v_j)) = \left[\prod_{F_i \in F_\alpha(v_j)}^{k}(1+\lambda w_i) - 1\right]/\lambda$; $I_h^l = \max_{v_j \in V} s(w_j)$.

We briefly described the fulfillment of the MoFNN as follows.

Step 1. We classified all samples of the data (5) (x_h^z, t_h) into two groups: A_1 is year with normal temperatures ($C_h^t=1$), A_2 is year with abnormal temperatures ($C_h^t=-1$). This classification provides vector with elements C_h^t.

Step 2. We trained two two-layer networks: $Y(x_h^z)$ (number of hidden neurons and delays are 7 and 2, respectively). The vector x_h^z was network's input. The vector C_h^t was network's target. We defined fuzzy sets A_j, (A_1 is year with normal temperatures, A_2 is year with abnormal temperatures) with membership function $\mu_j(s)$ based on aforementioned two-layer networks $Y(x_h^z), j=\overline{1..2}$.

Step 3. We trained MoFNN $s(o, z)$ based on the data (5). We elaborated optimization algorithm o (if $o = 1$ then optimization algorithm is ALO; if $o=2$ then optimization algorithm is Levenberg-Marquardt; if $o = 3$ then optimization algorithm is proposed modified multi-dimensional ALO algorithm). This step provides neural networks $F_{jq}(x_h^z)$ which forecast average monthly ambient temperature $Ir_{jq} = F_{jq}(x_h^z)$. We defined two agent's subcultures s_j based on aforementioned two-layer networks $F_{jq}(x_h^z)$.

If-then rules are defined as:

Π_j: IF x_h^z is A_j THEN $I_h = Fes\ (F_{jq}(x_h^z))$, (12)

We briefly described simulation of the trained MoFNN $\forall c \in [1215, 1226]$ as follows.

Step 1. Aggregation antecedents of the rules (12) maps input data x_c^z into their membership functions and matches data with conditions of rules. These mappings are then activating the k rule, which indicates the k year's temperature mode and k agent's subcultures – s_k, $k = \overline{1..2}$.

Step 2. According the k mode the modified fuzzy neuronet (generated base on the data x_d^z, where $d = \overline{1..c-1}$) forecasts average monthly ambient temperature $I_h = Fes(F_{kq}(x_c^z))$ as a result of multi-agent interaction of subculture s_k.

The fuzzy-possibilistic algorithm allows for the forecasting of average monthly ambient temperature in an intelligent manner, so as to take into account the responses of all agents based on fuzzy measures and the fuzzy integral.

3.4. Simulation Results of an Ambient Temperature Forecasting Model on the Basis of a Modified Fuzzy Neural Net

To demonstrate the advantages of the MoFNN in forecasting of the average monthly ambient temperature, the numerical examples from the previous subsections 3.1 - 3.3 are revisited using the software [9]. There the three MoFNN $s(o, z)$ were fulfilled based on the training set of the data (5), $t = \overline{1...1214}$. We trained the first MoFNN $s(3, z)$ using modified multi-dimensional ALO ($o = 3$). In order to obtain statistical results, we perform 120 modified multi-dimensional ALO and ALO runs with following parameters: S=50 (we use 50 ants and ant lions), $T = 110$ (we terminate at the end of 110 iterations), the dimension is d_h=4*h+3 $\in$

$\{D_{min} = d_1 = 7, D_{max} = d_{10} = 43\}$. The vector $f(best(d_h)) = (2.5, 2.1, 1.5, 1.1, 0.8, 1.2,$ 1.7, 2.4, 2.7, 3.1) demonstrates that only one set of MoFNN3 *(3, 1)* architecture with $d_5 = 23$ can achieves the fitness function (6) above 0.9 over data (5), $t = \overline{1215...1226}$. The MoFNN *s(3, z)* have three agents of each subculture s_k. We generated the aforementioned agents as the two-layered network with five hidden neurons. All the MoFNN *s(o, z)* have the same architecture. Forecast accuracies of the MoFNN *s(o, z)* we evaluated as the fitness function (6). Table 3 demonstrates comparison of the results between the MoFNN *s(o, z)*. Comparisons between temperature forecasting models MoFNN *s(o, z)* demonstrate that the MoFNN *s(3,1)* is definitely more accurate. Figure 1a) demonstrates that MoFNN *s(3,1)* has definitely more convergence speed than MoFNN *s(1,1)* or *s(1,2)* in the average monthly ambient temperature forecasting. The generation of initial personal best and best positions of a swarm based on Nguyen-Widrow method [23] provides an area of an optimum of a network's architecture at initial iteration and speed up the convergence process. In Figure 1b) the ineffectiveness of the *s(2,z)* can be seen.

The performance of the modified fuzzy neuronet generated by proposed algorithm is superior to the same one trained by ALO or Levenberg-Marquardt algorithm (Table 3, Figure 8).

This chapter presents a MoFNN for average monthly ambient temperature forecasting. Simulation comparison results for average monthly ambient temperature forecasting demonstrate the effectiveness of the MoFNN generated by proposed modified multi-dimensional ALO algorithm as compared with the same ones generated by ALO or Levenberg-Marquardt algorithm.

Table 3. The results of MoFNN at an average monthly ambient temperature forecasting

The MoFNN *s(o, z)*	*s(1,1)*	*s(1,2)*	*s(2,1)*	*s(2,2)*	*s(3,1)*	*s(3,2)*
The function (7) over the data (6), $t = \overline{1...1226}$	0.8	1.7	2.8	3.4	0.7	1.5

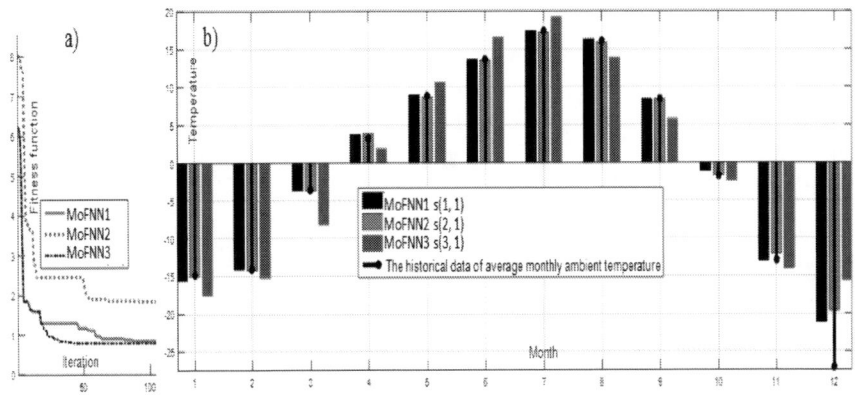

Figure 8. a) The mean convergence curves b) The results of MoFNN.

4. PHOTOVOLTAIC ARRAY CONTROL SYSTEM ON THE BASIS OF A MODIFIED FUZZY NEURAL NET

The Republic of Khakassia is one of the most perspective regions for development of solar power system in Russian Federation. The annual average of the solar insolation at the site of Khakassia is about 1450 kWh/sq.m [25]. That exceeds values of the European part of the Russian Federation (about 1200-1450 kWh/sq.m). But the PV systems aren't stable due to complex dynamics of the solar irradiance variations. Therefore, maximum power point tracking controllers have an important role in solar plant. We consider a non-linear MPPT problem for PV systems. PV system is non-linear and commonly suffers from restrictions imposed by sudden fluctuations in the solar irradiance level. Within the research literature, a whole array of differing MPPT algorithms has been proposed [13]. Among them, the perturbation & observation and incremental conductance algorithms are the most common due to simplicity and easy implementation. But controllers based on P&O, or IC algorithm for PV systems have slow response times to changing reference commands, take considerable time to settle down from oscillating around the target reference state, must often be designed by hand. Furthermore, the PV system control model should be robust to different environmental conditions, in order to reliably generate maximum power. Therefore, automatic intelligent algorithms such as fuzzy neural networks are promising alternatives [14].

The real-life PV systems have complex dynamic due to random variation of the system parameters and fluctuation of the solar irradiance. Thus, neural-network-based solutions have been proposed to approximate this complex dynamic [13]. But the neural network needs to become more adaptive. Adaptive behavior can be enabled by modifying the network into a recurrent neural network with fuzzy units. This forms the motivation for the development of a PV system control model on the basis of a modified fuzzy neural net. Compared to existing fuzzy neural nets, including ANFIS [21], the MoFNN includes recurrent neural networks and fuzzy units. The function approximation capabilities of a recurrent neural net are exploited to approximate a membership function. It is important to emphasize that a modified fuzzy neural net is improvement of a Multi-Agent Adaptive Fuzzy Neuronet [16].

4.1. The Photovoltaic System's Control Based on the Modified Fuzzy Neural Net

We exploited the function approximation capabilities of a MoFNN to approximate a nonlinear control law of PV system. We consider the development of an effective maximum PV power point tracking algorithm on the basis of a MoFNN. The proposed modified fuzzy neural net is capable of handling uncertainties in both the PV system parameters and in the environment.

We design and simulate in Octave environment 20 kW PV module by implementing following mathematical models of electrical characteristics. The open-circuit voltage is the extreme voltage offered from a PV cell at zero current. We calculate the open-circuit voltage as follows

$$V = NKT/Q \ln (I_L - I_o)/I_o + 1, \tag{13}$$

where V is the open-circuit voltage, N is diode ideality constant, K is the Boltzmann constant (1.381*10^-23 J/K), T is temperature in Kelvin, Q is electron charge (1.602*10^-19 c), I_L is the light-generated current same as Iph (A), and I_o is the saturation diode current (A).

We calculate the light-generated current as follows

$$I_L = G/G_{ref}(I_{Lref} + a_{Isc}(T_c - T_{c\,ref})), \tag{14}$$

where G is the radiation (W/m2), G_{ref} is the radiation under standard condition (1000 W/m2), I_{Lref} is the photoelectric current under standard condition (0.15 A), $T_{c\,ref}$ is module temperature under standard condition (298 K), a_{Isc} is the temperature coefficient of the short-circuit current (A/K) = 0.0065/K, I_L is the light-generated current.

We calculate the reverse saturation current as follows

$$I_o = I_{or} * (T/T_{ref})^3 \exp((QE_g/KN)*(1/T_r - 1/T)) \tag{15}$$

where $I_{or} = Ish/\exp^{(Vocn/NVtn)}$ is the saturation current, I_o is the reverse saturation current, N is the ideality factor 1.5, and E_g is the band gap for silicon 1.10 eV.

We calculate the short-circuit current as follows

$$I_{sh} = I_L - I_o(\exp(Q*(V-IR_S)/NKT) - 1) \tag{16}$$

This PV module provides 20 kW under standard condition (irradiance is 1000 W/m2, temperature is 20 C). This Octave model uses a duty cycle that generates the required voltage.

4.2. The PV Array Control System On The Basis Of a MoFNN

The MoFNN is trained based on the data

$$Z^i = (x^1 = (Ir^i, V^i, P^i, dI/dV^i);\ s^i = (\Delta Ir^i, dI/dV^i), D^i), \tag{17}$$

where $i \in \{1,..., 10^6\}$, I and V represent the current and voltage respectively, D^i is the duty cycle of boost converter, dI and dV represent (respectively) the current error and voltage error before and after the increment, Ir represents the solar irradiance, $\Delta\ Ir^i = Ir^i_0 - Ir^i_1$, Ir^i_0 is the irradiance before the increment, Ir^i_1 is the irradiance after the increment, P – the PV system power; x^i – input signal of MoFNN ; D^i – control signal. Data (17) have a training set of $8*10^5$ examples, and a test set of $2*10^5$ examples.

We briefly described fulfillment of the MoFNN as follows.

Step 1. All samples of the data (17) s^i were classified into two groups according to speed of the PV system conditions change: A_1 is sudden change ($C^{i1} = 1$), A_2 is smooth change ($C^{i2}=-1$). This classification generates vector with elements C^i.

Step 2. We trained two-layer network: $Y(s^i)$ (number of hidden neurons is 2). The vector s^i was network's input. The vector C^i was network's target. We formed membership function $\mu_j(s)$ based on the two-layer network $Y(s^i)$ as follows

$$\mu_1(s^i) = \begin{cases} Y(s^i), if\ Y(s^i) \geq 0 \\ 0, if\ Y(s^i) < 0 \end{cases}, \quad \mu_2(s^i) = \begin{cases} |Y(s^i)|, if\ Y(s^i) < 0 \\ 0, if\ Y(s^i) \geq 0 \end{cases} \qquad (18)$$

This step provides the fuzzy sets A_j, (A_1 is sudden change of the PV system conditions, A_2 is smooth change of the PV system conditions) with membership function $\mu_j(s)$, $j=\overline{1..2}$.

Step 3. We created the MoFNN based on the data (17). The MoFNN includes two recurrent neural networks F_j (number of delays is 2), $j=\overline{1..2}$. The MoFNN architecture's parameters (number of nodes in hidden layer, corresponded weights and biases) have been coded into particles X. The dimension component of particle X is $d_h = 12*h+2 \in \{D_{min} = d_1 = 14, D_{max} = d_{10} = 122\}$. To make the PV system control become adaptive, it needs to have some idea of how the actual PV system behavior differs from its expected behavior, so that the recurrent neural network F_j can recalibrate its behavior intelligently during run time, and try to eliminate the constant tracking error. We give the recurrent neural network $F_j(\mu_j(s),x)$ an extra input $\mu_j(s)$ which corresponds to the value of membership function $\mu_j(s)$. This input signal of the recurrent neural networks $F_j(\mu_j(s),x)$ will give useful feedback for providing the maximum PV power during the dynamically changing PV system conditions. This control approach does provide a more intelligent algorithm of generating the control signal u on the basis of a MoFNN. We evaluated the fitness function as follows:

$$f(D, u) = (1/H) \sum_{l=1}^{H} |D - u|. \tag{19}$$

where H is number of evaluated samples. We used modified multi-dimensional ALO as optimization algorithm. We presented modified multi-dimensional ALO in subsection 3.2. We used function (19) as a fitness function for the modified multi-dimensional ALO. This step provides trained MoFNN $best(d_h)$ which generate the control signal $u(best(d_h))$ – best solution X created by the modified multi-dimensional ALO.

If-then rules are defined as:

$$\Pi_j: \text{IF } x \text{ is } A_j \text{ THEN } u = F_j(^\mu{}_j(s),x), j = \overline{1..2}. \tag{20}$$

Simulation of the trained MoFNN briefly can be described as follows.

Step 1. Aggregation antecedents of the rules (20) maps input data x into their membership functions and matches data with conditions of rules. These mappings are then activate the k rule, which indicates the k PV system mode and correspondent k recurrent neural network $F_k(\mu_j(s),x)$, $k \in \overline{1..2}$.

Step 2. According the k mode the correspondent k recurrent neural network $F_k(\mu_j(s),x)$ (trained based on the data (17)) generates the control signal $u = F_j(\mu_j(s),x)$.

4.3. Simulations and Results

We revisited the numerical examples from the previous subsections 4.1 and 4.2 in order to illustrate the benefits of the proposed photovoltaic system control model based on a modified fuzzy neural net. All the simulations for this study are implemented in Octave.

Figure 9 demonstrates the solar irradiance during the simulation time. For the purpose of this simulation study, four solar irradiance scenarios were adopted:

1. From time = 0 sec to 0.4 sec graph demonstrates slow variable shadow cast by an obstacle, which causes a smooth change of irradiation;
2. From time = 0.5 sec to 1 sec graph demonstrates smooth and steady decline in solar irradiance which simulates a cloud covering;
3. From time = 1.1 sec to 2.1 sec irradiation changes to the exact target values with a smooth change;
4. From time = 2.1 sec to 2.5 sec graph demonstrates sudden change in irradiation, from sunshine conditions.

We fulfilled the MoFNN based on the training set of the data (17). We trained the MoFNN using modified multi-dimensional ALO. Due to obtain statistical results, we perform 120 modified multi-dimensional ALO runs with following parameters: $n = 50$ (we use 50 ants and ant lions), $T=100$ (we terminate at the end of 100 iterations), the dimension is $d_h = 12*h+2 \in \{D_{min} = d_1 = 14, D_{max} = d_{10} = 122\}$. The vector $f(best(d_h)) = $ (4.2e-3, 3.7e-4, 1.5e-5, 1.1e-5, 2.5e-6, 1.1e-7, 3.4e-8, 4.2,e-7 1.7e-5, 2.4e-4) demonstrates that only one set of MoFNN architecture with $d_7 = 86$ can achieves the fitness function (19) above 4e-8 over data (17).

This MoFNN includes two recurrent neuronets $F_k(\mu_k(s), x)$, $k = \overline{1..2}$. The aforementioned recurrent neuronets are the two-layered networks with seven hidden neurons. In this comparison study, the performance of the proposed PV system control model on the basis of a MoFNN is compared against the standard model with the PID controller (based on P&O or IC algorithm), under the same conditions. Figures 10 and 11 demonstrate the simulation results.

The proposed PV system control model is more robust and provides more power (Figures 10 and 11) in comparison with the control models with the PID controller (based on P&O, or the IC algorithm). Figure 10 demonstrates the misjudgment phenomenon for the P&O algorithm when solar irradiance continuously increases (time $t \in T = [0.3$ s, 0.4 s$] \cup [0.8$ s, 1 s$] \cup [1.7$ s, 2.1 s$]$). In such situations, the proposed PV system control model - which is based on a fuzzy modified neural net - produces on average 8.6% more energy than does the case of the standard model, which is based on a perturbation and observation algorithm (100% * ($\sum_{t \in T}(P_{MFNN}{}^t - P_{P\&O}{}^t)/P_{P\&O}{}^t / \sum_{t \in T} 1 = 8.6\%$, where P_{MFNN} is energy provided by proposed PV system control model based on a modified fuzzy neural net, $P_{P\&O}$ P_{MFNN} is energy provided by standard model based on P&O algorithm).

According to Figure 11, the response time using the IC algorithm is not better than the one using the proposed algorithm in the first 0.5 second. This means that the IC algorithm which creates the control signal within the transient mode is the overshoot. From time=2.2 s to 3 s the PV system energy providing by the control model with the PID controller based on the IC algorithm drops to zero.

During time $t \in [1.1$ s, 1.3 s$] \cup [1.5$ s, $1.7] \cup [2.2$ s, 3 s$]$ the PID controller based on the IC algorithm generates a huge numerical value of the control signal (value of control signal $u \in [-5.0706e+32; 5.6385e+33]$ (Figure 12)) as a result of sudden fluctuations in the solar irradiance, while the proposed PV system control model provided the maximum PV power (Figure 11).

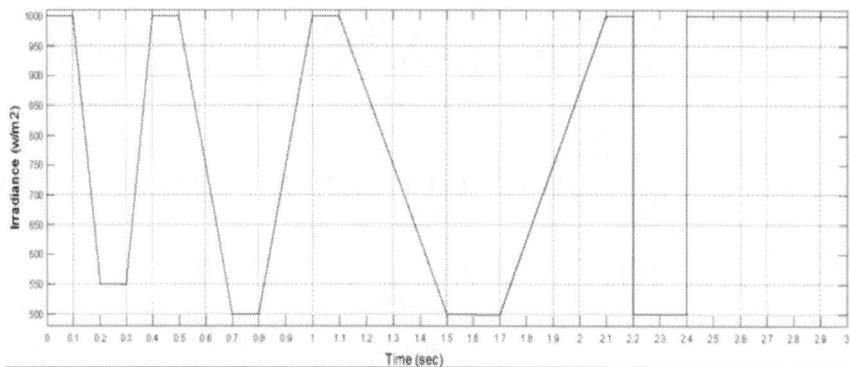

Figure 9. Plot of solar irradiance.

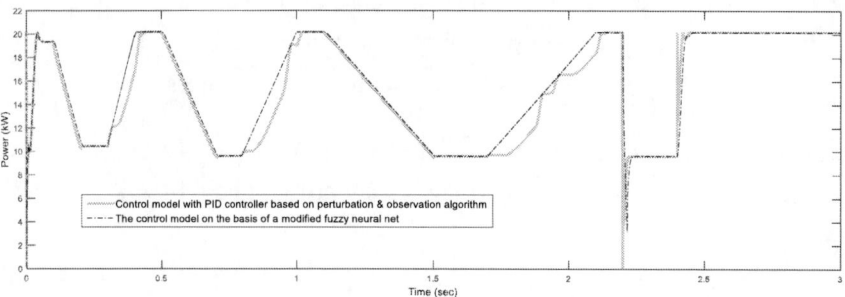

Figure 10. Plot of the PV system power provided by control model with PID controller based on P&O algorithm and the control model on the basis of a MoFNN respectively.

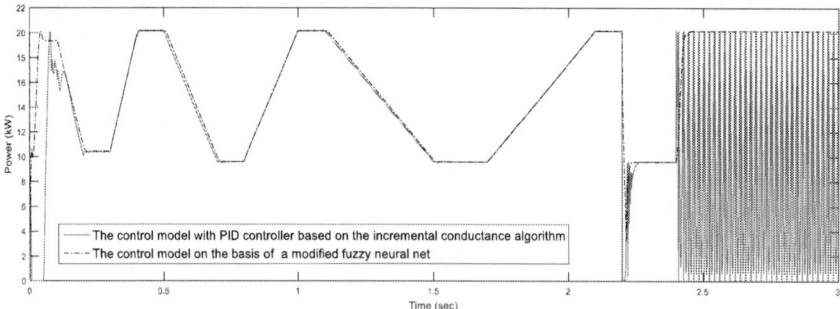

Figure 11. Plot of the PV system power provided by the control model with PID controller based on the IC algorithm and the control model on the basis of a MoFNN respectively.

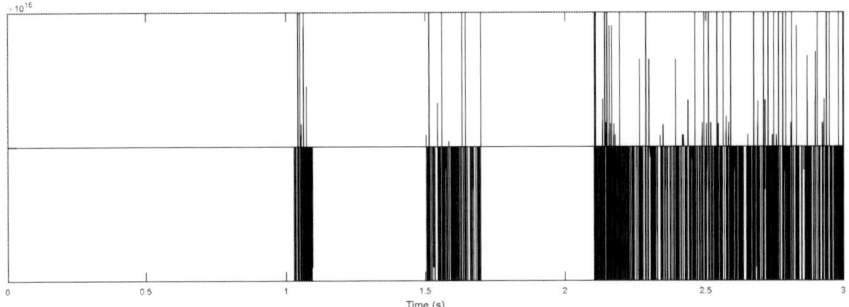

Figure 12. Plot of the control signal provided by the PID controller base using the Incremental Conductance algorithm.

The MoFNN provides a more suitable approach to the MPPT problem, with the pointing accuracy. Extensive simulation studies on the Octave model have been carried out on different initial conditions, different disturbance profiles, and variation in photovoltaic system and solar irradiation level parameters. The simulation results demonstrate that consistent performance has been achieved for the proposed PV system control model on the basis of a MoFNN with good stability and robustness as compared to a standard model with a PID controller.

The simulation results demonstrate that the PV system control model on the basis of a MoFNN is robust to PV system uncertainties. Unlike popular approaches to nonlinear control, a MoFNN is used to approximate the control law and not the system nonlinearities, which makes it suitable over a wide range of nonlinearities. Compared to standard MPPT algorithms, including P&O and IC, the PV system control model on the basis of a MoFNN produces good response time, low overshoot, and, in general, good performance. In summary, simulation comparison results for a PV system demonstrate the effectiveness of the PV system control model on the basis of a MoFNN as compared with the standard model with a PID controller (based on P&O, or IC algorithm). It is our contention that the proposed modified fuzzy neural net produces a competitive alternative algorithm to neural networks and PID controllers.

5. Sizing of a Photovoltaic System with Battery on the Basis of the Modified Fuzzy Neuronet

The effectiveness of a solar plant is depending on annual average of the solar insolation at the site of a solar plant, size and type of PV modules and storage capacity. Within the research literature, a whole array of differing sizing algorithms for a photovoltaic system has been proposed [2]. The sizing algorithms of a photovoltaic system are classified as intuitive algorithms, numerical algorithms, and analytical algorithms. The intuitive algorithms aren't provide effictievness and reliability. The numerical algorithms require a long time series of solar insolation. Many of the analytical algorithms use the concept of the system's reliability or the complementary term: loss of load probability. Maleki A and Askarzadeh A presented an intelligence algorithm for PV system size optimization based [3]. Optimal sizing of PV systems is a very complex task which requires the fulillment of mathematical models for the photovoltaic system's components as well as usage of global optimization algorithms. We solves the task of an optimum photovoltaic system's sizing based on modified fuzzy neural net. We collected a database of total solar radiation data, meteorological parameters of ambient temperature at the site of Khakassia from 01 April 2016 through 01 September 2019. The Khakassia locates in the south part of Siberia (91.4° of longitude East, 53.7° of latitude North and 246 m of altitude), Russian Federation. We determined the optimal PV sizing coefficients at the site of Khakassia by minimizing the objective function on the basis of modified fuzzy neural net. The optimal sizing coefficients define the number and type of PV modules and the battery's capacity. Finally, we performed a comparison experiment between a modified fuzzy neural net solution and an ANFIS and PSO in sizing of a PV system.

5.1. Mathematical Model of the Photovoltaic System Components

The PV system includes a photovoltaic module, a solar regulator, a battery bank and an inverter. Figure 13 demonstrates a scheme of a basic PV system.

The total rate of radiation G_C striking a PV system on a clear day is calculated as follows:

$$G_c = Ae^{-km}\left(\cos b \cos(\varphi_s - \varphi_c)\sin S + \sin b \cos S + C\left(\frac{1+\cos S}{2}\right) + p(\sin b + C)\left(\frac{1-\cos S}{2}\right)\right) \quad (21)$$

where m is the air mass, β is the altitude angle, φ_S is the solar azimuth angle, φ_C is the photovoltaic module azimuth angle, p is the reflection factor, S is the photovoltaic module tilt angle, C is the sky diffuse factor, A and k are parameters related to the Julian day number.

The solar regulator's is dimensioned according to its input current, given as

$$I^M = \frac{N_{pv} \times P_{pv}}{h_{rg} \times U \times N_{pvs}}, \quad (22)$$

where N_{pv} is the total number of PV modules, N_{pvs} is the number of PV modules in series, η_{rg} (%) is the efficiency of the regulator and U is the nominal system operating voltage (V).

5.2. Objective Function

The main goal is to derive a maximum power from a PV module. We grouped the variables of the proposed objective function into two classes: dependent variables: $[b, m, G_C, I_{SC}, I_0, T_c]$ and associated control variables: $[N_S, N_P, d, S, f_C]$. We identified objective function as follows:

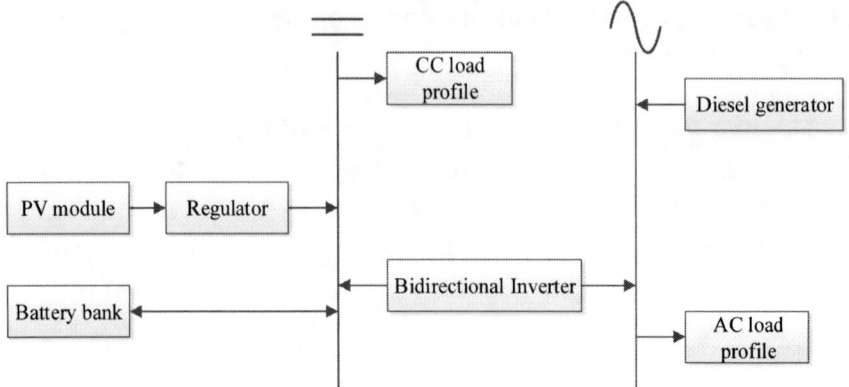

Figure 13. The scheme of a basic PV system.

$$P_{PV}^{i(max)}(t, S_{opt}) = f(T_c, V^M, m, S, f_C, b, w, G_C, I_0) = (N_S V^C) I_{n+1}^M$$
$$= V^M (I_n^M * \left[1 - \frac{1}{1 + y(T_c)*(1 - m(V^M, T_c, I^M))*\P(T_c)} \right] +$$ (23)
$$\frac{y(T_c)*(m, f_S, b, S, f_C)*m(V^M, T_c, I^M) + g(V^M)}{1 + y(T_c)*(1 - m(V^M, T_c, I^M))*\P(T_c)}),$$

where $P_{pv}^{i(max)}(t, S_{opt})$ is the maximum power of the photovoltaic module at optimal tilt angle S_{opt} and hour t during a day no. i, $y(T_c)$, $m(V^M, T_c, I^M)$, $\P(T_c)$, $e(m, f_S, b, S, f_C)$ and $g(V^M)$ are nonlinear functions.

We presented the parametric limits for (23) in the following way:

$$\begin{aligned} d^{min} &< d < d^{max} \Rightarrow 1 \le d \le 365 \\ S^{min} &< S < S^{max} \Rightarrow 0 \le S \le 80 \\ f^{min} &< f < f^{max} \Rightarrow -45 \le f \le 45 \end{aligned}$$ (24)

We defined equality constraint as follows:

$$g(U, X) = V_{oc} - 184.0293 * \frac{N_s V^C}{T_c} = 0$$ (25)

We identified the parametric limits for (23) due to the following aspects:

1. If $s = 0°$ the module becomes horizontal then it produces power. If $\Sigma = 90$ the module turns vertical then it doesn't provide power. Therefor the optimal tilt angle belonged $[0°, 80°]$.
2. On south-west the solar azimuth angle has negative positive value. On south-east the solar azimuth angle is positive. Hence the solar azimuth angle belonged $[-45°, 45°]$.

We evaluated the total power delivered from the PV array to the battery bank throughout day i and hour t is as follows:

$$P_{re}^i(t) = N_{pv} * P_{re}^{i(\max)}(t, S_{opt}) \tag{26}$$

where N_{pv} is the total number of photovoltaic modules compose the array.
We evaluating the DC/AC inverter input power in the following way:

$$P_L^i(t) = \frac{P_{load}^i(t)}{x_{inv}} \tag{27}$$

where x_{inv} is efficiency of the inverter, $P_{load}^i(t)$ is the power exhausted by the load at hour t of day i ($P_{load}^i(t)$ evaluated at the first step of the optimal sizing). In accordance with (26) and (27) capacity of the battery is evaluated.

- If $P_{re}^i(t) = P_L^i(t)$ then the battery capacity stays unchanged.
- If $P_{re}^i(t) > P_L^i(t)$ then the Pv power $P_B^i(t) = P_{re}^i(t) - P_L^i(t)$ is consumed to charge the battery.

We evaluate the new battery capacity in the following way

$$C^i(t) = C^i(t-1) + \frac{P_B^i(t) * Dt * x_{bat}}{V_{Bus}} \tag{28}$$

where $1 \leq t \leq 24$, $C^i(t)$, $C^i(t-1)$ is the available battery capacity (Ah) of day i at hour t and $t-1$, respectively, $x_{bat} = 100\%$ during discharging, $x_{bat} = 80\%$ is the battery round-trip efficiency during charging and V_{Bus} is the DC bus voltage, $P_B^i(t)$ is the battery input/output power, Δt is the simulation time step (in this research $\Delta t = 1h$ because the data of load and solar irradiance have same step). The storage capacity has the following constraints:

$$C_{min} \leq C^i(t) \leq C_{max}, \qquad (29)$$

where C_{max}, C_{min} are the maximum and minimum acceptable storage capacities, $C_{min} = DOD * C_n$, where C_n is the nominal battery's capacity.

The number of photovoltaic modules connected in series is given as follows:

$$n_{pv}^s = \frac{V_{bus}}{V_{MP}}, \qquad (30)$$

where V_{bus} is the nominal DC bus voltage.

The number of batteries is given as follows:

$$n_b^s = \frac{V_{bus}}{V_b} \qquad (31)$$

where V_b is the nominal voltage of each individual battery.

The number of battery chargers is given as follows:

$$N_{ch} = \frac{N_{pv} * P_m}{P_{mc}} \qquad (32)$$

where P_m is the photovoltaic module's maximum power under standard test conditions, P_{mc} is the battery charger's coefficient.

We identified the total photovoltaic system cost function as follows:

$$\min\{J(u)\} = \min\{C_c(u) + C_m(u)\} \tag{33}$$

where $C_c(u)$ and $C_m(u)$ are the total capital and maintenance cost functions respectively, $u=(u_1, u_2)$ is a vector of the cost independent variables, u_1 is the total number of photovoltaic modules, u_2 is the total number of batteries.

We evaluated the optimal value of u before the total number of battery chargers is defined. Thus, we fulfilled the multi-objective optimization by minimizing the total cost function. The total cost function includes individual system cost devices capital cost and 20 year round maintenance cost.

We identified life cycle cost objective function as follows:

$$J(u) = \sum \frac{\sum_{i=1}^{N_{PV}} i(C_{PVi} + 20 \times M_{PVi})}{L.T_{PV}} +$$
$$+ \sum \frac{\sum_{j=1}^{N_{BAT}} j \times C_{BATj}(1 + y_{BATj} + M_{BATj} \times (20 - y_{BATj}))}{L.T_{BAT}}$$
$$+ \sum \frac{\sum_{l=1}^{N_{CH}} l \times C_{CHl}(1 + y_{NCHl} + M_{NCHl} \times (20 - y_{NCHl}))}{L.T_{CH}}$$
$$+ \sum \frac{C_{Inv}(1 + y_{Inv} + M_{Inv} \times (20 - y_{Inv}))}{L.T_{Inv}} \tag{34}$$

where $N_{PV} \geq 0$, $N_{BAT} \geq 0$, $L.T_{PV}, L.T_{BAT}, L.T_{CH}, L.T_{INV}$ are the year life time for a photovoltaic module, a battery, a battery charger and an inverter respectively; $u = [N_{PV}, N_{BAT}]$, C_{PV} and C_{BAT} are the capital costs of one photovoltaic module, and battery, respectively; M_{PV} and M_{BAT} are the maintenance costs per year of one photovoltaic module and battery, respectively; C_{ch} is the capital cost of one battery charger; y_{ch}, y_{inv} are the estimated numbers of the battery charger and DC/AC inverter changings during the 20 year system lifetime (y_{ch} and y_{inv} are constants, $y_{ch} = y_{inv} = 4$); C_{inv} is the capital cost of an inverter, y_{BAT} is the estimated number of battery changings during the 20 year system maintenance; M_{ch} and M_{inv} are maintenance costs per year of one battery charger and DC/AC inverter, respectively.

For each component we estimated the maintenance cost per year as 1% of the associated capital cost. We evaluated the total optimal number of PV modules N_{PV} and the total optimal number of batteries N_{BAT} by minimizing (34). Then, we evaluated the number of parallel strings n_{pv}^p and the number of batteries connected in parallel n_b^p as follows:

$$n_{pv}^p = \frac{N_{pv}}{n_{pv}^s}, \qquad (35)$$

$$n_b^p = \frac{N_{BAT}}{n_b^s}. \qquad (36)$$

Thus, for the photovoltaic system constituents we provided the optimal number and optimal configuration. We evaluated the various photovoltaic systems' configurations and the optimal cost of each configuration based on (34). Afterwards we identified the minimum cost and the associated photovoltaic systems' configuration. The photovoltaic systems' configuration includes the optimal number and optimal type of the PV system components.

In order to obtain the optimal PV system's cost we elaborated the modified fuzzy neural net and the PSO.

5.3. Sizing of Photovoltaic System

The main objective of this research is to define the optimal sizing coefficient (n_{pv}^{opt}, n_b^{opt}) by maximizing (23) and minimizing (34). In order to achieve the aforementioned goal we elaborated the modified fuzzy neural net and the PSO. Figure 14 presents the steps of the proposed PV sizing optimization algorithm based on the PSO (O is the PSO) or the modified fuzzy neural net (O is the modified fuzzy neural net).

PSO is a stochastic optimization algorithm based on a quasi-random search of the solution space. The key advantages between the PSO and other global optimization algorithms are the easy implementation and fast convergence. Therefore PSO has received growing implementation in PV applications. The PSO describes the optimization process by the position and velocity vectors. Figure 15 demonstrates flow chart diagram of a pure PSO.

PSO provides global optimization of complex and irregular spaces. In order to determine the optimal sizing coefficient we elaborated the PSO. A database (containing the technical characteristics of commercially available system devices along with their corresponding per unit capital and maintenance costs) provided input data for photovoltaic sizing optimization algorithms based on the PSO. We considered different types of PV modules, batteries with different nominal capacities, etc. We fulfilled the PV sizing PSO based on the data

$$(CP_{it}, G_{it}, T_{it}, N_{PV}^{j}, N_{BAT}^{j}), \qquad (37)$$

where G_i is the data of solar irradiance, N_{PV}^{j} is the total optimal number of photovoltaic modules, N_{BAT}^{j} is the total optimal number of batteries, T_i is ambient temperature, CP_i is distribution of the consumer power requirements during a day, $i=\overline{1..1248}$, $t=\overline{1..24}$. Number of samples for each type of photovoltaic modules ($j=\overline{1..3}$) is 24* 1248 = 29472. Figure 2 presents the steps of the proposed PV sizing optimization algorithm based on the PSO (O is the PSO).

The photovoltaic sizing optimization algorithm based on the PSO provides data

$$(CP_{it}, G_{it}, N_{PV}^{a}, N_{BAT}^{a}, T_{it}, PVmax_{it}^{a}, J^{a}), \qquad (38)$$

where N_{PV}^{a} is the total optimal number of photovoltaic modules, N_{BAT}^{a} is the total optimal number of batteries, $PVmax_{it}^{a}$ is the maximum power from a photovoltaic module $a=\overline{1..3}$, J^{a} is life cycle cost.

Unlike existing single-agent and team-search methods in the proposed modified fuzzy neural net the agents collaborate through knowledge of other agents during the simulation. This simulation we fulfilled based on real data (38). We fulfilled the agents of the modified fuzzy neural net based on two-layered recurrent networks. The modified fuzzy neural net implemented offline learning. The algorithm of the multi-agent search uses the neuroevolution (Figure 16).

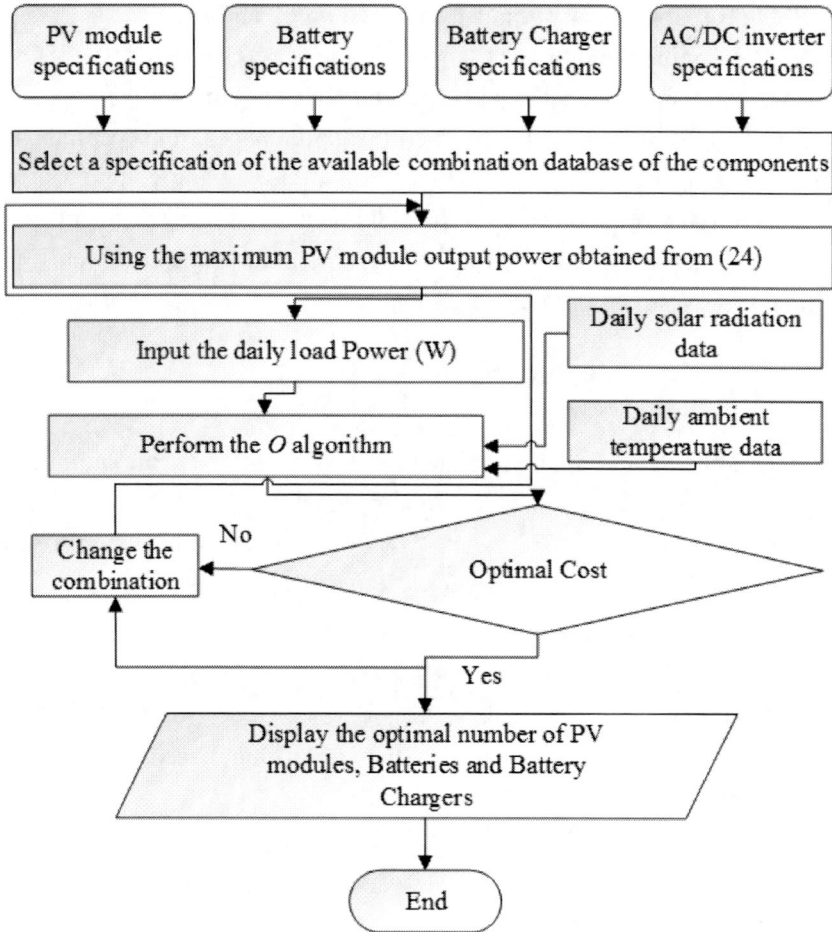

Figure 14. Flow chart of the proposed photovoltaic sizing optimization algorithm.

Photovoltaic Applications on the Basis of Modified Fuzzy Neural Net 49

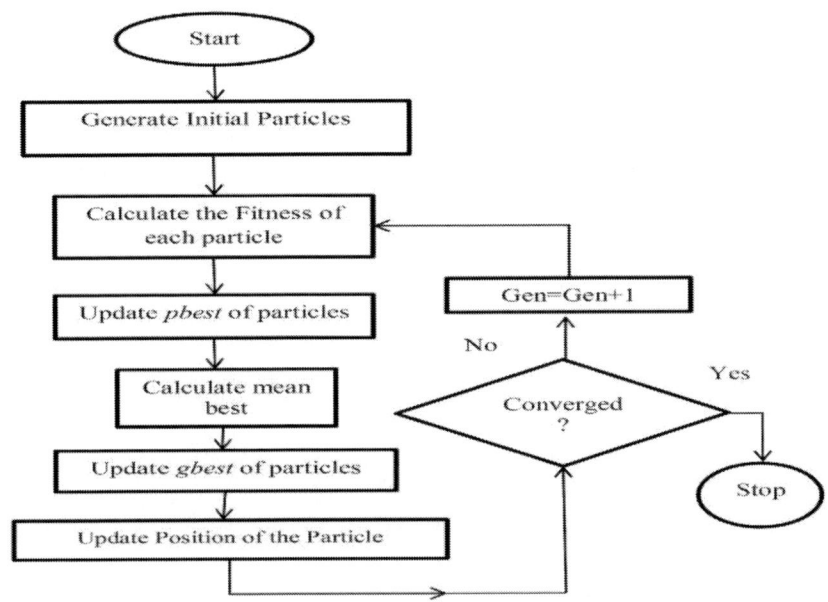

Figure 15. Flow chart diagram of a PSO.

```
Input: generations; maxTimesteps
population ← InitializePopulation()  g ← 0  t ← 0
loop while g < generations do
  subcultures ← FormSubcultures(population)
      while t < maxTimesteps do
  for each agent in subculture S_k do
          s_i^t ← ReceiveAgentInputs()
          f_ja(s_i^t) ← GetResponse(agent; s)
          v_a ← TakeAction(f_ja(s_i^t))
          agent.memory ← UpdateMemory({s_i^t; f_ja(s_i^t); v_a})
  end for
  for each agent in subculture S_k do
          if Acceptable(agent.memory) then
              for all observer in subculture do
                      Train(observer, agent.memory)
              end for
          end if
  end for
              end for
              t ← t + 1
          end while
          population ← SelectAndReproduce(population)
          g ← g + 1
end while
end loop
```

Figure 16. Algorithm of the multi-agent search.

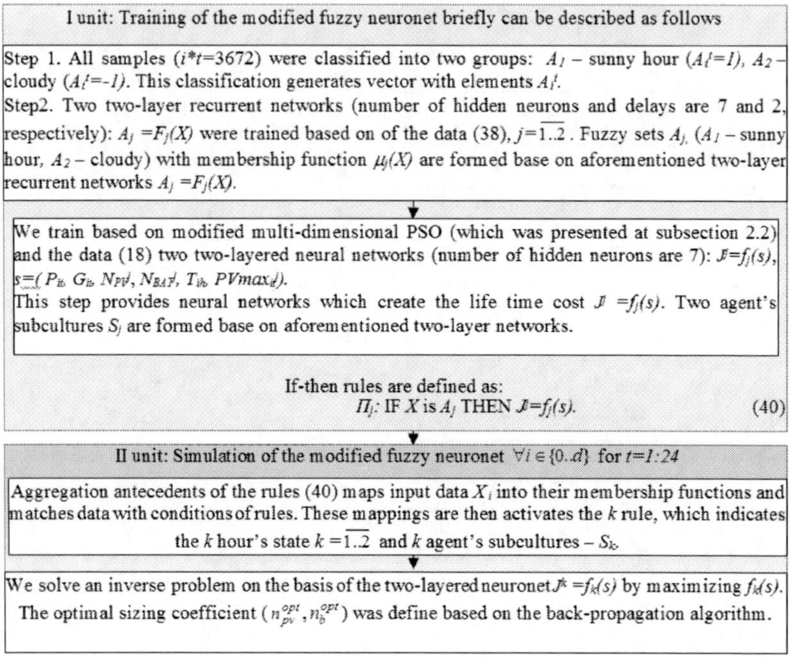

I unit: Training of the modified fuzzy neuronet briefly can be described as follows
Step 1. All samples ($i*t=3672$) were classified into two groups: A_1 – sunny hour ($A_i^t=1$), A_2 – cloudy ($A_i^t=-1$). This classification generates vector with elements A_i^t. Step 2. Two two-layer recurrent networks (number of hidden neurons and delays are 7 and 2, respectively): $A_j = F_j(X)$ were trained based on of the data (38), $j=\overline{1..2}$. Fuzzy sets A_j, (A_1 – sunny hour, A_2 – cloudy) with membership function $\mu_j(X)$ are formed base on aforementioned two-layer recurrent networks $A_j = F_j(X)$.
↓
We train based on modified multi-dimensional PSO (which was presented at subsection 2.2) and the data (18) two two-layered neural networks (number of hidden neurons are 7): $J^j = f_j(s)$, $s = (P_L, G_L, N_{PV}^j, N_{BAT}^j, T_{ih}, PVmax_d^j)$. This step provides neural networks which create the life time cost $J^j = f_j(s)$. Two agent's subcultures S_j are formed base on aforementioned two-layer networks.
If-then rules are defined as: Π_j: IF X is A_j THEN $J^j = f_j(s)$.   (40)
II unit: Simulation of the modified fuzzy neuronet $\forall i \in \{0..d\}$ for $t=1:24$
Aggregation antecedents of the rules (40) maps input data X_i into their membership functions and matches data with conditions of rules. These mappings are then activates the k rule, which indicates the k hour's state $k = \overline{1..2}$ and k agent's subcultures – S_k.
↓
We solve an inverse problem on the basis of the two-layered neuronet $J^k = f_k(s)$ by maximizing $f_k(s)$. The optimal sizing coefficient (n_{pv}^{opt}, n_b^{opt}) was define based on the back-propagation algorithm.

Figure 17. The modified fuzzy neural net.

In this chapter, we exploited the function approximation capabilities of a modified fuzzy neural net to approximate a life cycle cost objective function based on data (38). Figure 14 presents the steps of the proposed PV sizing optimization algorithm on the basis of the trained modified fuzzy neural net (the O is the trained modified fuzzy neural net). We briefly described fulfilling of the modified fuzzy neural net by Figure 17.

5.4. Results

To illustrate the experimental validations of the modified fuzzy neural net in finding the best configuration of the photovoltaic system, we revised the numerical comparison results by usage the author's software [9]. The data (37) includes the solar irradiation (Figure 18) and the ambient temperature collected at the site of Khakassia (91.4° of longitude East, 53.7° of latitude North and 246 m of altitude).

Photovoltaic Applications on the Basis of Modified Fuzzy Neural Net 51

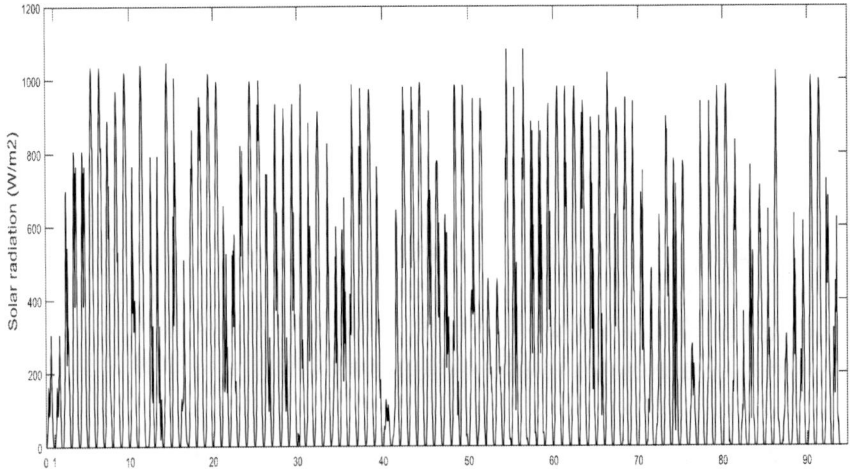

Figure 18. The data of the measured solar insolation at the site of Khakassia, Russian Federation.

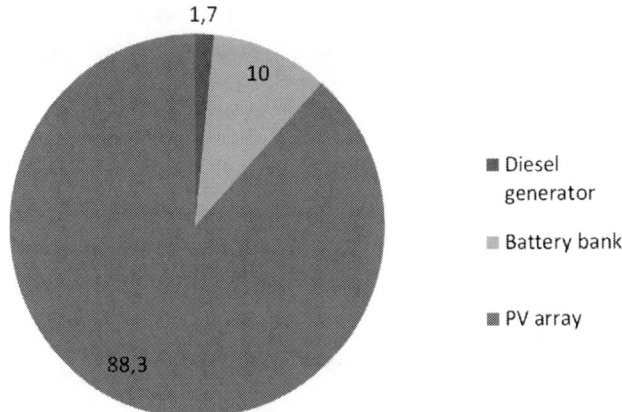

Figure 19. Percentage of energy provided by PV array, diesel generator and battery at the site of Khakassia, Russian Federation.

The total energy demand per day for the load is equal to 4.5 kW h/day. We implemented the proposed photovoltaic sizing optimization algorithm based on the author's software [9].

Figure 19 demonstrates the percentage of energy provided by photovoltaic, diesel generator, and battery over a year.

As can be seen, the solar power plants can be applied at the site of Khakassia (91.4° of longitude East, 53.7° of latitude North and 246 m of altitude) with a high potential for solar energy.

We performed a comparison between the modified fuzzy neural net solution and the PSO solution. Table 4 briefly summarizes the results of the optimum photovoltaic system sizing by the modified fuzzy neural net and the PSO. Table 4 demonstrates that the modified fuzzy neural net provides a better solution than does a PSO. Table 4 demonstrates that modified fuzzy neural net provided the optimal cost which is 7.019 $/wh (1951.3 $/year). The comparison results demonstrate that the modified fuzzy neural net provided optimum solar and battery ratings. The performance of the modified fuzzy neuronet in optimum photovoltaic system sizing is superior to the PSO.

The result of optimum photovoltaic system sizing at the site of Khakassia (Russian Federation) demonstrates that solar plants can be applied for this location with a high potential for solar energy. Thus, usage of solar energy can be considered as a good alternative to a coal-fired power station at the site of Khakassia, Russian Federation. The utilization of the proposed algorithm can help to overcome some of the barriers that still limit the distribution of solar energy projects. It can be also used as a starting point (or as a support tool) to promote and design efficient solar energy projects.

Table 4. A comparison between the PSO solution and the modified fuzzy neural net solution

PV module type	The PSO solution						The modified fuzzy neuronet solution			
	Battery (Ah)	Charger (W)	N_{PV}	N_{Ch}	N_{Batt}	Cost ($/wh)	N_{PV}	N_{Batt}	N_{Ch}	Cost ($/wh)
CS5C-90	230	1152	12	1	13	8.018	12	11	1	7.019
Bpsx150	230	288	10	1	15	9.473	11	13	1	8.337
CS6P-200	100	1152	7	1	19	10.612	6	18	1	10.741

6. MAXIMUM PHOTOVOLTAIC SYSTEMS POWER POINT TRACKING ALGORITHM BASED ON MODIFIED PARTICLE SWARM OPTIMIZATION UNDER NON-UNIFORM IRRADIANCES

Under fast-changing non-uniform solar irradiance level multiple local maximum power points are observed but only one global maximum power point exists which lead to a dynamic optimization problem. This paper presents a maximum power point tracking algorithm of a photovoltaic array based on the modified particle swarm optimization. The modification of particle swarm optimization employs re-randomization and the particles around a maximum power point initialization. The simulation results demonstrate that the proposed algorithm provides stability and fast maximum power point tracking of a photovoltaic array, as compared to a classical Perturb and Observe or particle swarm optimization algorithms under fast-changing non-uniform solar irradiance level.

Solar energy is the most inexhaustible and environment friendly among all the clean and renewable energy resources. The main advantages of photovoltaics include its applicability in most regions of the world, both in industrial and household applications. The power generated by photovoltaic array dependents on temperature, solar irradiation level and shading. The power-voltage characteristic of partially shaded PV array have several peaks and conventional maximum power point tracking algorithm, such as Perturb and Observe, fails to track the global maximum power point under fast-changing non-uniform solar irradiance level [1]. There are many optimization algorithms implemented in MPPT controller [2–4]. Among those, in [3] particle swarm optimization to track the GMPP is presented. A pure PSO algorithm can work as MPPT and set a direct duty cycle [4]. A velocity equation provides the convergence of the particles to the GMPP. The PSO algorithm exhibits considerable potential, due to easy implementation, fast computation capability, and its ability to determine the GMPP irrespective of fast-changing non-uniform irradiation level of a PV system. One of the drawbacks of the PSO algorithm is that during slow changing in the solar irradiance level, the alteration of the control signal needs to be small in order to track the GMPP properly. The initialization of the swarm in PSO is a crucial issue affecting performance. This forms the motivation to initialize the particles around the maximum power point based on constant voltage algorithm approximation and interpolation polynomial in the Lagrange form.

Compared to the conventional PSO algorithm the proposed MPPT algorithm provides closeness of new operating points to the MPP and requires less iteration. The comparison simulation study of the proposed algorithm and classical Perturb and Observe or particle swarm optimization algorithms under fast-changing non-uniform solar irradiation level is elaborated. The simulation results demonstrate that the proposed algorithm provides stability and fast maximum power point tracking, as compared to a classical Perturb and Observe or particle swarm optimization algorithms under fast-changing non-uniform solar irradiation level.

6.1. Model of the Photovoltaic Module

The Shockley's simple "one diode" model describes the operating of a PV module [5]. Figure 20 demonstrates an equivalent circuit of this model.

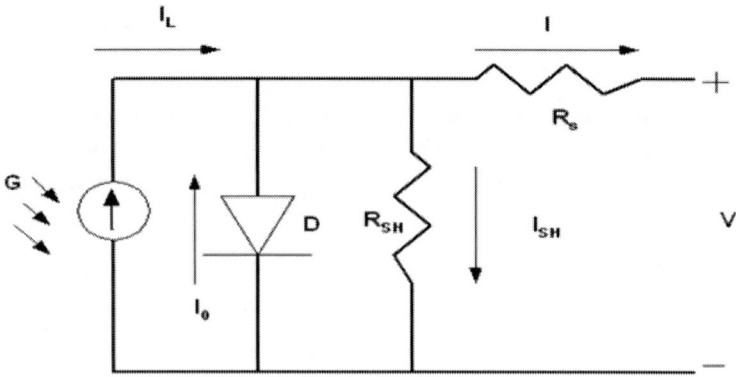

Figure 20. The equivalent circuit of the one diode model.

The corresponding current-voltage (I-V) characteristic expression is written as:

$$I = I_{ph} - I_o[exp(q(V+I \cdot R_s)/(N_{cs}gkT_c))-1] - (V+IR_s)/R_{sh}, \qquad (39)$$

where I – current supplied by the module (A), V – voltage at the terminals of the module (V), I_L – photocurrent (A), I_o – inverse saturation current (A), R_s – series resistance (ohm), R_{sh} – shunt resistance (ohm), q – charge of the electron, $k = 1.381\ E\text{-}23$ – Bolzmann's constant (J/K), g – diode quality factor (normally between 1 and 2), N_{cs} – number of cells in series, T_c – effective temperature of the cells (Kelvin).

A model of the 250-W photovoltaic array in the Octave environment under non-uniform solar irradiance level was fulfilled. This Octave model implements a PV array built of three series connected PV modules with the bypass diodes neutralizing negative current under a partial shading condition.

Table 5 demonstrates the parameters of the PV module.

Figure 21 demonstrates photovoltaic module I–V and P-V characteristics.

Table 5. Parameters OF PV MODULE

Parameter	Value
Open circuit voltage Voc (V)	12.64
Short-circuit current Isc (A)	10.32
Temperature coefficient of Isc (%/deg.C)	0.063701
Temperature coefficient of Voc (%/deg.C)	- 0.33969
Maximum power (W)	83.2824
Cells per module (Ncell)	20
Shunt resistance (ohm)	82.1161
Series resistance (ohm)	0.098625
Light-generated current (A)	8.6307
Diode saturation current (A)	1.4176e-10
Diode ideality factor	0.99132

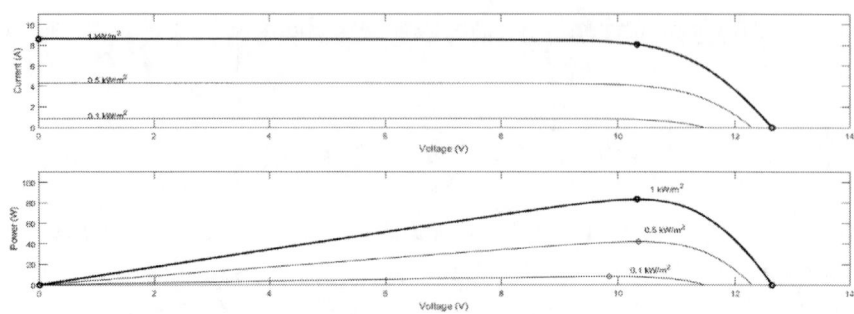

Figure 22. The photovoltaic module I–V and P-V characteristics.

6.2. MPPT Algorithms under Fast-Changing Non-Uniform Solar Irradiation Level

PSO is a stochastic optimization algorithm based on a quasi-random search of the solution space. The key advantages between the PSO and other global optimization algorithms are the easy implementation and fast convergence. Therefore PSO has received growing implementation in MPPT PV systems. PSO describes the optimization process by the position and velocity vectors. Figure 23 demonstrates flow chart diagram of a pure PSO.

The MPPT controller based on P&O algorithm adjusts the voltage by a small amount and measures power. It is referred to as a hill climbing algorithm, because it depends on the rise of the curve of power against voltage below the maximum power point, and the fall above that point. Figure 24 demonstrates flow chart diagram of Perturb and Observe MPPT algorithm.

Perturb and Observe MPPT algorithm is the most common due to simplicity and easy implementation. But controllers based on Perturb and Observe MPPT algorithm for PV systems have slow response times to changing reference commands, take considerable time to settle down from oscillating around the target reference state, and must often be designed by hand.

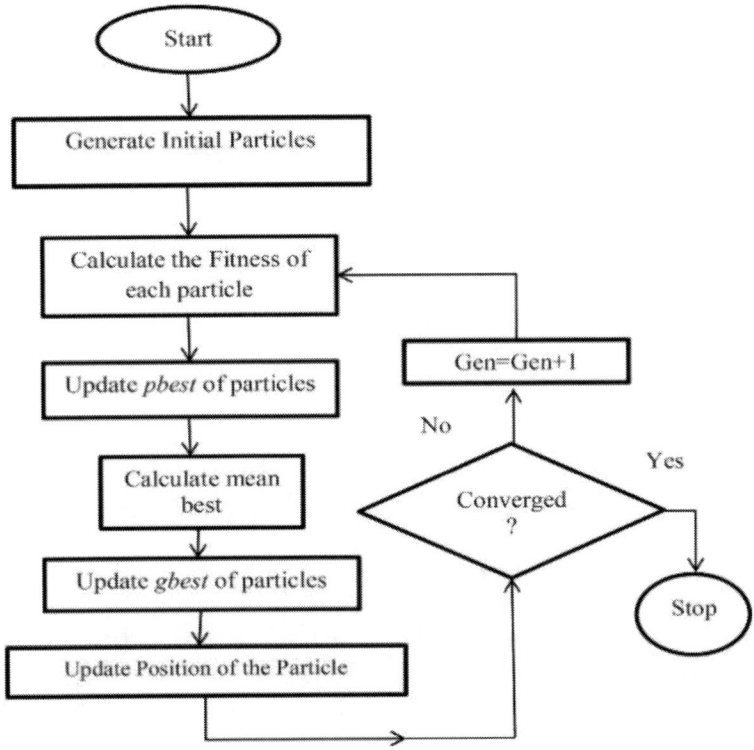

Figure 23. Flow chart diagram of a pure PSO.

If solar irradiance continuously increases then the misjudgment phenomenon of Perturb and Observe MPPT algorithm occurs. In such cases Perturb and Observe MPPT algorithm moves operating point in direction which opposites the global maximum point.

For the proposed MPPT algorithm, the control signal – PV voltage have been coded into position of particle X.

The modified PSO process at j-th iteration is represent by the following characteristics:

$p_{X,j}$: the personal best (pbest) position of particle X;

$x_{X,j}$: the position of particle X;

$v_{X,j}$: the velocity of particle X;

e_j: the evaporation rate;

g: Global Best position of swarm;

w: inertia;
c_2 : cognitive weight;
c_3: social weight;
E: the base value of the evaporation rate.

We the calculated w, c_2, c_3 based on the constricted PSO formula [6] and their values are 0.73, 1.5 and 1.5, respectively.

We evaluated the fitness function of the modified PSO at iteration as power of the PV array as follows:.

$$f(x_{X,j}) = P_j = V_j I_j, \qquad (40)$$

where I_j – current supplied by the PV array at j-th iteration (A), V_j – voltage of the PV array at j-th iteration (V), P_j denotes the power of the PV array at j-th iteration.

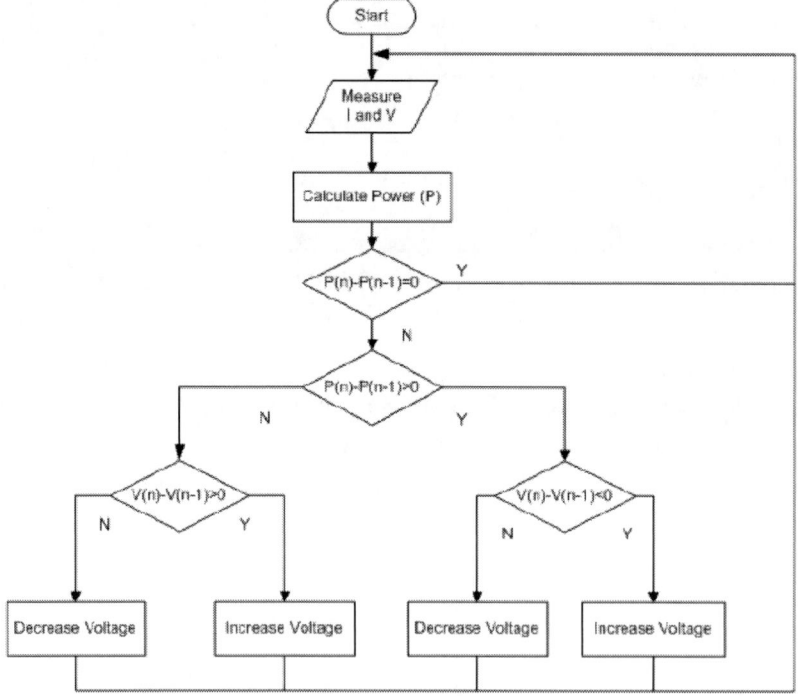

Figure 24. Flow chart diagram of Perturb and Observe MPPT algorithm.

The proposed algorithm includes procedure L based on constant voltage algorithm approximation and interpolation polynomial in the Lagrange form. This procedure L estimates x – voltage at MPP based on following equation:

$$L(x_i, y, i = \overline{0..3}) = \frac{\prod_{i=1}^{3}(x-x_i)}{\prod_{i=1}^{3}(x_0-x_i)} y_0 + \ldots + \frac{\prod_{i=0}^{2}(x-x_i)}{\prod_{i=0}^{2}(x_3-x_i)} y_3, \qquad (41)$$

where x_0 represents the voltage $V_0 = 0$ of the short circuit current, x_3 represents the open circuit voltage which provided by the PV array data sheet, x_i is voltage and the corresponding value y_i represents the duty cycle ($i = \overline{0..3}$, $y_0 = 1$, $y_3 = 0$).

The task of PV array MPPT after a quick change of the global maximum power point represents the dynamic optimization problem. Because of that proposed MPPT algorithm elaborates growth of the evaporation rate and re-randomization during fitness function updating in order to locate the global maximum.

We expressed the proposed MPPT algorithm as follows:

Step 0. We initialize the particle's positions $x_{1,0}$ and $x_{2,0}$ as present and previous (stored at the end of the preceding cycle) voltage values, respectively, $j=0$.

Step 1. For $\forall X \in \{1, 4..S/3\}$ do

We initialize the particle's positions at fixed, equidistant points, which positioned around the MPP: $x_{X,j} = x - v_{X,j}$, $x_{X+1,j} = x$, $x_{X+2,j} = x + v_{X+2,j}$ where x is the voltage at MPP which estimated based on procedure (3) $L(x_{X-i,j}, y_i, i = \overline{1..2})$. Therefore, initial values of the particle's positions $x_{X,j}$ set close to the MPP.

If $j > 0$ Then $S=2*S$, $r=0$ go to step 3 End If.
End For.

Step 2. For $\forall X \in \{1, ...S\}$ do
Initialize the particle's best known position
If $f(x_{X,0}) > f(p_{X-1,0})$ then Do: $p_{X,0} = x_{X,0}$ End If.
If $f(x_{X,0}) > \max(f(p_{X-1,0}), \max_{1 \leq k < X}(f(x_{k,0})))$

Then Do $g = x_{X,0}$. End If.
End For.
Step 3. While a convergence criterion is not met do:
For $\forall j \in \{1,T\}$ do:
For $\forall X \in \{1,S\}$ do:
Pick random numbers: $r_1, r_2 \sim U(0,1)$
Update the particle's velocity:
$v_{X,j+1} = wv_{X,j} + c_1 r_1 (p_{X,j} - x_{X,j}) + c_2 r_2 (g - x_{X,j})$.
If $X_{min} \leq v_{X,j+1} \leq X_{max}$ then $x_{X,j+1} = x_{X,j} + v_{X,j+1}$ else $x_{X,j+1} = U(X_{min}, X_{max}) + x_{X,j}$ End If.
If $f(x_{X,j}) > f(p_{X,j-1})$ then Do: $p_{X,j} = x_{X,j}$ else $p_{X,j} = p_{X,j-1}$ End If.
If $f(x_{X,j}) > \max(f(p_{X,j-1}), \max_{1 \leq k < X}(f(x_{k,j})))$
Then Do $g = x_{X,j}$. End If.

$$\text{If} \quad (F_j > 0 \quad or \quad \sum_{i=1}^{S} f(x_{i,j}) < 0.75 \sum_{i=1,k=1}^{S,7} f(x_{i,j-k}))$$

Then $F_{j+1} = F_{j+1}$. End If.
In the case of a relatively rapid decline in the average power for all particles (when the power in the j-th iteration is lower than 75% of the power in one of the past 7 iterations), the rate of evaporation grows exponentially.
If $f(x_{X,j}) > e_j p_{X,j-1}$ Then $F_{j+1} = 0$, $e_j = E^{1+F_j}$,
$p_{X,j} = e_j p_{X,j-1}, r = r + 1$ End If
If $r \geq 3$ Then $x_{1,0} = p_{X,0}, x_{2,0} = g, S = S/2$, go to step1 End If.
End For. End For.

The advantages of the proposed algorithm include a faster convergence obtained by procedure (41) than for pure PSO MPPT algorithm and a greater probability of finding the GMPP under partial shading condition than for a classical Perturb and Observe MPPT algorithm.

6.3. Results

To illustrate the benefits of the proposed MPPT algorithm based on the modified particle swarm optimization, the numerical examples from the previous sections 6.1 and 6.2 are revisited. All the simulations of the 250-W PV array under fast-changing non-uniform solar irradiance level are carried out based on Octave model.

Figure 25 demonstrates the solar irradiance during the simulation time. In this simulation study we adopted three solar irradiance scenarios:

1. From time = 0 sec to 1 sec graph presents fast change in solar irradiance from partial shading to sunshine cases;
2. From time = 1.1 sec to 2 sec graph presents smooth and steady decline in solar irradiation which simulates a cloud covering;
3. From time = 2.1 sec to 3 sec solar irradiance changes smoothly which simulates slow shadow because of an obstacle.

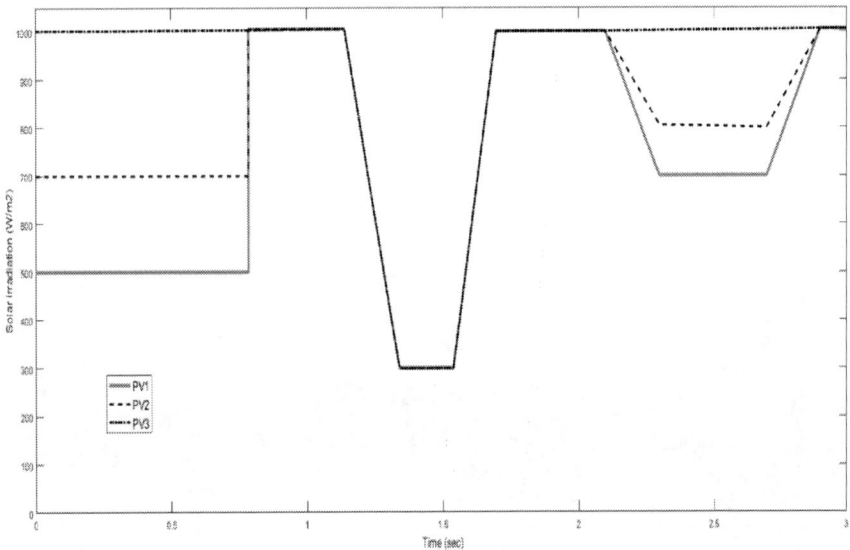

Figure 25. Plot of solar irradiance.

Usually solar plant starts generating power in the morning under self-shading condition (Figure 26).

In order to capture more direct irradiance, the solar modules are usually kept at higher tilt angles. But with such increased tilt, the length of shadow casted to the row behind increases. This leads to self-shading of PV modules which happens in solar plants especially at winter.

Shading by soiling is common case at solar plants. The term 'soiling' is used to describe the accumulation of snow, dirt, dust, leaves, pollen, and bird droppings on PV panels. The performance of a PV module decreases by surface soiling, and the PV power loss increases with an increase in the quantity of soil on the PV module. Thus, the surface soiling of PV modules leads to a significant loses in energy produced by a PV array.

The condition becomes even worse in some situations such as snowfall on PV modules (Figure 27). According to statistic studies the power loss can vary from 10% to 70% due to partial shading [7]. The first scenario implemented the cases of self-shading and soiling shading.

Table 6 demonstrates the swarm parameters of pure PSO and the proposed algorithm.

Figure 26. The solar plant in the morning under self-shading condition.

Photovoltaic Applications on the Basis of Modified Fuzzy Neural Net 63

Figure 27. The solar plant under soiling shading condition.

Table 6. Parameters of swarm of pure PSO and the proposed MPPT algorithm

Parameter	PSO	Proposed MPPT algorithm
Swarm size	10	3
Particle dimension	1	1
Base value of evaporation rate	0.9	0.9

In the case of first scenario the objective function has three local maxima (Figure 28). The classical Perturb and Observe MPPT algorithm is trapped in the zone of the local maximum area (Figure 28 demonstrates three operating points, marked as asterisk, provided by Perturb and Observe MPPT algorithm). The pure PSO algorithm in such case, just activates the initialization procedure for particles (Figure 28 demonstrates three operating points, marked as square, provided by pure PSO algorithm). The proposed MPPT algorithm in such a case, after taking the sudden cumulative power drop of the PV array into account, activates the re-randomization procedure for half consecutive particles which provides quick convergence to a global maximum (Figure 28 demonstrates three operating points marked as circle provided by proposed MPPT algorithm).

The second scenario, which occurred at 1.2 sec, simulates a cloud that covers PV array at an appropriate speed to trigger the re-randomization mechanism of the proposed MPPT algorithm. This does not happen at a uniform irradiance change that occurs over all cells of PV array (Figure 29 demonstrates three operating points marked as circle provided by proposed MPPT algorithm). Figure 29 demonstrates that the re-randomization mechanism is not necessary because the proposed MPPT algorithm copes precisely with solar irradiation changes. The usage of re-randomization in second scenario would decrease the speed of convergence to the global maximum. The pure PSO algorithm in such case, just activates the initialization procedure for particles (Figure 29 demonstrates three operating points, marked as square, provided by pure PSO algorithm). In the second scenario the solar irradiance continuously increases, which occurred from 1.5 sec to 1.7 sec, and figure 29 demonstrates the misjudgment phenomenon of Perturb and Observe MPPT algorithm. In such situation P&O MPPT algorithm moves operating point in direction which opposites the global maximum point (Figure 29 demonstrates three operating points marked as asterisk provided by proposed MPPT algorithm).

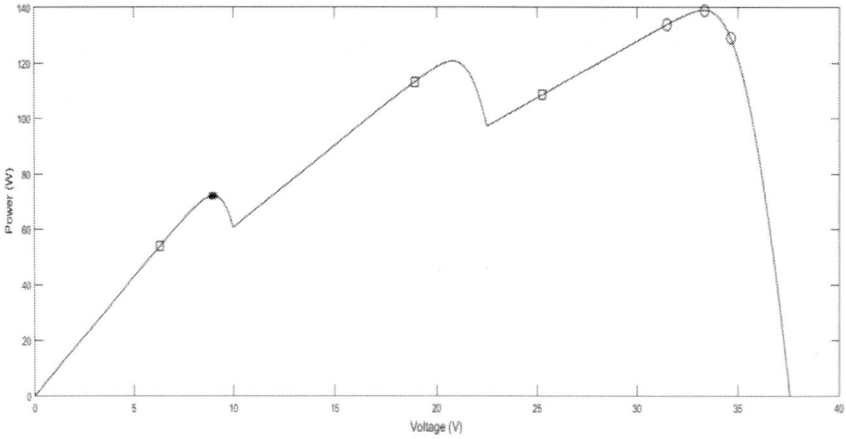

Figure 28. The numerical results of the MPPT algorithms in the first scenario.

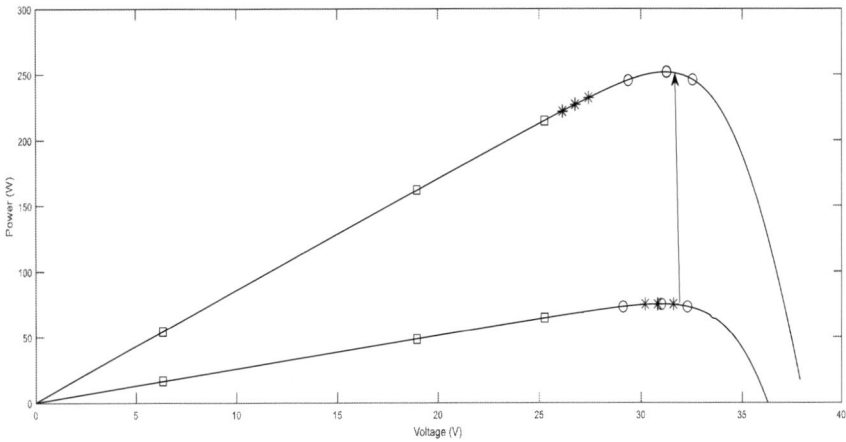

Figure 29. The numerical results of the MPPT algorithms in the second scenario.

The third scenario, which occurred at 2.1 sec, simulates a shadow cast by an obstacle, covering cells of PV array non-uniformly. Subsequently, the objective function has three local maxima (Figure 30). Even a slower rate of irradiance, compared to the second scenario, the proposed MPPT algorithm triggers the re-randomization (Figure 30). As a result, the proposed MPPT algorithm quickly converges to a global maximum (Figure 30 demonstrates three operating points marked as circle provided by proposed MPPT algorithm). The classical Perturb and Observe MPPT algorithm is ineffective to finding of the global maximum in such conditions. Figure 30 demonstrates the misjudgment phenomenon for the Perturb and Observe algorithm when solar irradiance continuously increases. (Figure 30 demonstrates three operating points, marked as square, provided by Perturb and Observe MPPT algorithm) The pure PSO algorithm in such a case, just activates the initialization procedure for particles (Figure 30 demonstrates three operating points, marked as square, provided by pure PSO algorithm).

Table 7 summarizes the simulation results of the proposed algorithm, a classical Perturb and Observe and particle swarm optimization algorithms under fast-changing non-uniform solar irradiance level accordingly three solar irradiance scenarios.

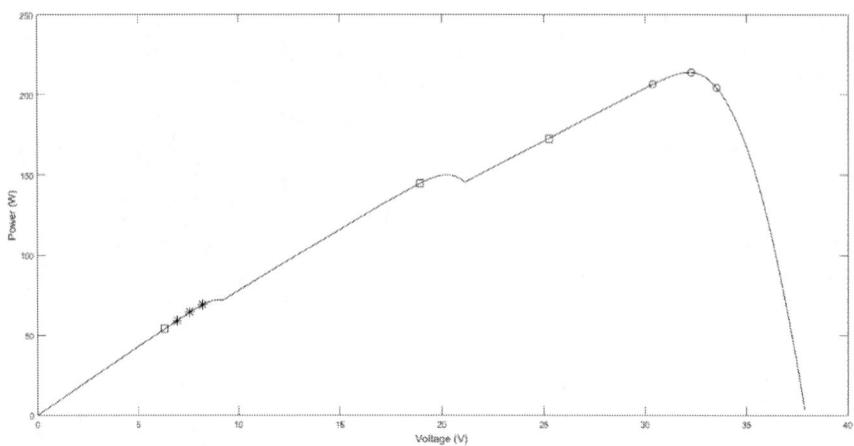

Figure 30. The numerical results of the MPPT algorithms in the third scenario.

Table 7. The average power of PV array provided by pure PSO, P&O MPPT algorithm and the proposed MPPT algorithm for the first and third scenarios

Ir of PV1 (W/m²)	Ir of PV2 (W/m²)	Ir of PV3 (W/m²)	P&O (W)	PSO (W)	Proposed algorithm (W)	Maximum power of PV array (W)
1000	700	500	70	135	210	215
1000	900	800	72	82	135	138

It is clear that the PV array power provided by the proposed MPPT algorithm is above 99.5% under all solar irradiance scenarios. The simulation results demonstrate that the proposed algorithm provides stability and quick convergence to a global maximum, as compared to a classical Perturb and Observe or particle swarm optimization algorithms under fast-changing non-uniform solar irradiation level.

This paper proposes a maximum power point tracking algorithm of a photovoltaic array based on the modified particle swarm optimization. The modification of particle swarm optimization employs re-randomization and the particles around the maximum power point initialization. This efficient initialization of particles provides avoiding both unnecessary and redundant searching and a situation in which the area of swarm search doesn't include global maximum. As a result, this significantly reduces the time that wasted by searching GMPP in the wrong area. Thus substantially enhancing the system's tracking speed, while also reducing the steady-state oscillation (practically to zero) once the GMPP is located. This is a huge improvement upon the conventional PSO algorithm which provides the new operating point too far from the MPP and requires much iteration to reach GMPP. To illustrate the benefits of the proposed MPPT algorithm based on the modified particle swarm optimization, the numerical simulation is carried out. The simulation results demonstrate that the proposed algorithm provides stability, faster response rate and fast maximum power point tracking, as compared to a classical Perturb and Observe or particle swarm optimization algorithms under fast-changing non-uniform solar irradiation level. The utilization of the proposed MPPT algorithm can help to overcome some of the technical barriers that still limit the PV array power provided by the MPPT. It can be also used as a starting point or a support tool to promote and design efficient solar energy projects faster.

CONCLUSION

The real-life PV systems have complex non-linear dynamic due to random variation of the PV system parameters and fluctuation of the solar irradiance. Thus, neural-network-based solutions have been proposed to approximate this complex dynamic. But the neural net needs to become more adaptive. Modification of the neural net into a recurrent neuronet with fuzzy units provides adaptive behavior. This forms the motivation for the development of intelligent models for sizing, parameters forecasting and control of a photovoltaic system on the basis of a modified fuzzy neural net. Compared to existing fuzzy neuronets, including ANFIS [21], the modified fuzzy neural net includes recurrent neuronets and fuzzy units. The function approximation capabilities of a recurrent neural net are exploited to approximate a membership function. The algorithm of the agent's interaction uses a fuzzy-possibilistic method. The optimum architecture of the modified fuzzy neural net was generated based on proposed algorithm which combines swarm and gradient optimization algorithms. A modified fuzzy neural net have the resilience to noise of the low similarity of the responses for patterns of the different PV system's state, and the ability to have similar responses for patterns contaminated with different intensities of noise. These two features provide the robustness of a modified fuzzy neural net to noise. The proposed modified fuzzy neural net is capable of handling uncertainties in both the PV system's parameters and in the environment.

This chapter presents the intelligent models for sizing, parameters forecasting and control of a photovoltaic system on the basis of a modified fuzzy neural net. The modified fuzzy neural net provides automatic fulfillment and modification of all proposed intelligent models.

The first proposed model on the basis of a modified fuzzy neural net provides a two days ahead hourly PV system power forecast under random perturbations. The PV system power forecasting is critically essential for planning effective transactions in power system operation, because it provides a safety of grid control. We fulfilled the MoFNN based on an extensive empirical database. We collected a database of the total power from a PV system, ambient temperature, meteorological parameters and insolation data in the south-eastern part of Siberia, Russian Federation at the site of Khakassia from 01 March 2016 through 01 July 2019. In order to train the effective MoFNN we developed the modified multi-dimensional PSO, in which the multi-dimensional PSO is combined with the Levenberg-Marquardt algorithm. The generation of initial position of a swarm based on Nguyen-Widrow method speed up the convergence process. The comparison simulation results leads us to the conclusion that proposed modified multi-dimensional PSO algorithm outperforms multi-dimensional PSO and Levenberg-Marquardt algorithm in training the effective MoFNN for the PV system power forecasting. The performance of the MoFNN trained by proposed algorithm is superior to the same one trained by multi-dimensional PSO or Levenberg-Marquardt algorithm, especially during fast variations of cloudiness. The analysis of the evolving errors demonstrates the potential of the MoFNN in the hourly PV array power forecasting.

The second proposed model on the basis of a modified fuzzy neural net is ambient temperature forecasting model under random perturbations. We generated the MoFNN architecture's parameters (an agent's number, a number of nodes in hidden layer, corresponded weights and biases) from the global optimum which provided by proposed modified multi-dimensional ALO algorithm. The multi-dimensional modification of an ALO provides an area of a global optimum of a modified fuzzy network's architecture, and then for a best solution of iteration we apply the Levenberg-Marquardt algorithm in order to speed up the convergence process. The generation of initial personal best and best positions of a swarm based on Nguyen-Widrow method [23] speeds up the convergence process at initial iteration. The simulation results demonstrate that proposed modified multi-dimensional ALO outperforms ALO and Levenberg-Marquardt algorithm in training the optimum MoFNN for average monthly ambient temperature forecasting.

The third proposed model on the basis of a modified fuzzy neural net is a photovoltaic system control model. Unlike popular approaches to nonlinear control, a modified fuzzy neural net approximated the control law of a PV system and not the PV system nonlinearities. According the photovoltaic system mode, the modified fuzzy neural net provides a maximum power point tracking under random perturbations. We carried out the extensive simulation studies on the PV system control models under different initial conditions, different disturbance profiles, and fluctuation in PV system and solar irradiation level. Compared to a classical control model with a PID controller based on perturbation & observation, or incremental conductance algorithm, the photovoltaic system control model on the basis of a modified fuzzy neural net produces good response time, low overshoot, robustness and in general good performance. It is our contention that the proposed modified fuzzy neural net produces a competitive alternative algorithm to Neural Networks and PID controllers.

The fourth proposed model on the basis of a modified fuzzy neural net provides the best configuration and optimal sizing coefficient of a photovoltaic system. We considered the combination of PV array and battery storage for this purpose. We developed the modified fuzzy neural net due to obtain the best configuration of the PV system and sizing of the PV system components. We defined the PV system power and life cycle cost as objective functions. In order to illustrate the benefits of the proposed photovoltaic sizing optimization algorithm based on the modified fuzzy neural net, we carried out numerical simulations. For this purpose we used the solar irradiation and the ambient temperature collected at the site of Khakassia (Russian Federation). This research solves the task of a photovoltaic system's sizing on the basis of a modified fuzzy neural net. The experimental results demonstrate that the modified fuzzy neural net provides a better solution than does a PSO. The comparison results demonstrate that the modified fuzzy neural net provided optimum solar and battery ratings. The performance of the modified fuzzy neuronet in optimum photovoltaic system sizing is superior to the PSO. The result of optimum photovoltaic system sizing at the site of Khakassia (Russian Federation) demonstrates that solar plants can be applied for this location with a high potential for solar energy. Thus, usage of solar energy can be considered as a good alternative to a coal-fired power station at the site of Khakassia, Russian Federation. The proposed PV system sizing algorithm can be also used as a starting point (or as a support tool) to promote and design efficient solar energy projects.

The photovoltaic applications of the modified fuzzy neural net for solving practical tasks revealed its experimental validation and following advantages:

- It supports the real time mode, low overshoot and competitive performance, as compared to classical algorithms;
- A trained modified fuzzy neural net effectively processes noisy data.

The fifth intelligent model on the basis of proposed modified PSO provides a maximum power point tracking of a partially shaded PV array. The proposed modified PSO employs re-randomization and the particles around the maximum power point initialization. Simulation comparison results demonstrate that compared to a PSO, the proposed modified PSO significantly reduces the computational time. Thus substantially enhancing the system's tracking speed, while also reducing the steady-state oscillation (practically to zero) once the GMPP is located. This is a huge improvement upon the conventional PSO algorithm which provides the new operating point too far from the MPP and requires more iteration to reach GMPP. Simulation comparison results demonstrate the competitive effectiveness, a maximum PV system power and real-time control speed of the modified fuzzy neural net as compared to a perturbation & observation algorithm and PSO in finding the global maximum power point of a partially shaded photovoltaic array.

The utilization of the proposed intelligent models can help to overcome some of the barriers that still limit the distribution of solar energy projects.

ACKNOWLEDGMENTS

The author wishes to thank Daniel Foty for valuable comments. The reported study was funded by Russian Foundation for basic research and Republic of Khakassia according to the research project № 19-48-190003.

REFERENCES

[1] Ayman, Youssefa., Mohammed, El-Telbanya. & Abdelhalim, Zekry. (2017). "The role of artificial intelligence in photovoltaic systems design and control: A review." *Renewable and Sustainable Energy Reviews, 78,* 72-79. doi:10.1016/j.rser.2017.04.046.
(Youssefa, El-Telbanya and Zekry 2017, 72-79).

[2] Shrestha, G. B. & Goel, L. (1998). "A study on optimal sizing of stand-alone photovoltaic stations." *IEEE Trans Energy Convers, 13*(4), 373-378. doi:10.1109/60.736323.
(Shrestha and Goel 1998, 373-378).

[3] Maleki, A. & Askarzadeh, A. (2014). "Comparative study of artificial intelligence techniques for sizing of a hydrogen-based stand-alone photovoltaic/wind hybrid system." *International journal of hydrogen energy, 39*(19), 9973-9984. doi:10.1016/j.ijhydene.2014.04.147.
(Maleki and Askarzadeh 2014, 9973-9984).

[4] Tavares, C. A. P., Leite, K. T. F., Suemitsu, W. I. & Bellar, M. D. (2009). "Performance evaluation of photovoltaic solar system with different MPPT methods." *Paper presented at IECON '09, 35th Annual Conference of IEEE.* Industrial Electronics, Porto, Portugal, November 3-5.
(Tavares 2009).

[5] Hua, C., Lin, J. & Shen, C. (1998). "Implementation of DSP-controlled photovoltaic system with peak power tracking." *IEEE Trans, Industrial Electronics., 45*, 99-107. doi:10.1109/41.661310.
(Hua, Lin and Shen 1998, 99-107)

[6] *Technology platform "Small-scale Distributed Power Generation."* 2013. Last modified March 11. http://www.e-apbe.ru/distributed_energy/

[7] Engel, E. A. (2013). "The method of constructing an effective system of information processing based on fuzzy-possibilistic algorithm." *Paper presented at the the 15th International Conference on Artificial Neural Networks* (Neuroinformatics-2013), Moscow, Russian Federation.
(Engel 2013).

[8] Engel, E. A. & Engel, N. E. (2019). "Photovoltaic System Control Model on the Basis of a Modified Fuzzy Neural Net" *Paper presented at the 19th International Conference Neuroinformatics*, 2019, Moscow, Russian Federation, October 7-11.
(Engel 2019)

[9] Engel, E. A. (2016). *"The intellectual system for forecasting of a non-linear technical object's state."* Certificate about State registration of computer programs. M.: Federal Service For Intellectual Property (Rospatent). № 2016663468.

[10] Engel, E. A. (2016). *The intellectual controller of a non-linear technical system.* Certificate about State registration of computer programs. M.: Federal Service For Intellectual Property (Rospatent). № 2016663467.

[11] Engel, E. A. & Engel, N. E. (2018). *The intelligent forecasting of power for a photovoltaic module.* M.: Federal Service for Intellectual Property (Rospatent), Certificate about State registration of computer programs №2018612282.

[12] Engel, E. A. & Engel, N. E. (2018). "Temperature Forecasting Based on the Multi-agent Adaptive Fuzzy Neuronet." Paper presented at the *18th International Conference Neuroinformatics*, 2018, Moscow, Russian Federation, October 8-12.
(Engel 2018)

[13] Tavares, C. A. P., Leite, K. T. F., Suemitsu, W. I. & Bellar, M. D. (2009). "Performance evaluation of PV solar system with different MPPT methods." Paper presented at IECON '09, *35th Annual Conference of IEEE*.
(Tavares 2009)

[14] Kumar, A., Chaudhary, P. & Rizwan, M. (2015). "Development of fuzzy logic based MPPT controller for PV system at varying meteorological parameters." *Paper presented at Annual IEEE India Conference (INDICON)*, New Delhi, India.
(Kumar 2015).

[15] Sobri, S., Koohi-Kamali, S. & Rahim, N. A. (2018). "Solar photovoltaic generation forecasting methods: A review." *Energy Convers. Manag., 156*, 459–497. https://umexpert.um.edu.my/file/publication/00003228_163877_75951.pdf.
(Sobri, Koohi-Kamali and Rahim 2018, 459-497)

[16] Ekaterina, A. Engel. (2018). "Modeling, Sizing and Control of a Photovoltaic System on the Basis of a Multi-Agent Adaptive Fuzzy

Neuronet" In: *Photovoltaic Systems: Design, Performance and Applications,* edited by Wassila Issaadi, Salim Issaadi, 139-180. Nova Science Publishers. ISBN: 978-1-53613-646-3. Series: Electrical Engineering Developments.
(Engel 2018, 139-180)

[17] Wassila, Issaadi., Malika, Mazouzi. & Salim, Issaadi. (2018). "The influence of the sampling frequency on various commands MPPT: Comparison and persepective" *International Journal of Solar Energy.* *Elsevier.* https://www.researchgate.net/publication/ 308899986_An_Improved_MPPT_Converter_ Using_Current_Compensation_Method_for_PV-Applications
(Issaadi 2018)

[18] Wassila. Issaadi. (2017). "Control of a photovoltaic system by Fuzzy Logic, comparative studies with conventional controls: results, improvements and perspectives." *International Journal of Intelligent Engineering Informatics,* *5*(3), 206-224. https://www.academia.edu/ 32752392/Control_of_a_photo voltaic_system_by_fuzzy_logic_comparative_studies_with_conventi onal_controls_results_improvements_and_perspectives.
(Issaadi 2017)

[19] Wassila, Issaadi. (2016). "An Improved MPPT Converter Using Current Compensation Method for PV-Applications." *International Journal of Renewable Energy Research, 6.3*, 894-913. https:// www.researchgate.net/publication/308899986_An_Improved_MPPT _Converter_Using_Current_Compensation_Method_for_PV-Applications. (Issaadi 2016).

[20] Wassila, Issaadi., Abdelkrim, Khireddine. & Salim, Issaadi. (2016). "Management of a base station of a mobile network using a photovoltaic system." *International Journal of Renewable & Sustainable Energy Reviews (Elsevier), 59,* 1570-1590. doi:10.1016/j.rser.2015.12.054.
(Issaadi 2016)

[21] Jang, J. S. R. (1993). "ANFIS: adaptive-network-based fuzzy inference system" IEEE *Transactions on Systems, Man and Cybernetics*, *23*(3), 665-685. doi:10.1109/21.256541.
(Jang 1993, 665-685)

[22] Serkan, Kiranyaz., Turker, Ince., Alper, Yildirim. & Moncef, Gabbouj. (2009). "Evolutionary artificial neural networks by multi-dimensional particle swarm optimization" *Neural Networks*, *22*(10), 1448-1462. doi.org/10.1016/j.neunet.2009.05.013.
(Kiranyaz et. al. 2009, 1448-1462)

[23] Derrick, Nguyen. & Bernard, Widrow. (1990). "Improving the learning speed of 2-layer neural net-works by choosing initial values of the adaptive weights." Paper presented at the International Joint Conference on Neural Networks, San Diego, California USA, June 17-21.
(Derrick 1990).

[24] Mirjalili, S. (2015). "The ant lion optimizer." *Advances in Engineering Software*, *83*, 80-98. doi :10.1016/j.advengsoft.2015.01.010
(Mirjalili 2015)

[25] BetaEnergy. (2019). "Solar irradiance of regions". Accessed June 12 2019. https://www.betaenergy.ru/insolation/abakan.

BIOGRAPHICAL SKETCH

Ekaterina A. Engel

Affiliation: Associate Professor, First Russian doctoral degree in computer sciences, Katanov State University of Khakassia, Abakan, Russian Federation

Education:

- 2009. Postdoctoral studies in Computer Science and Engineering, University of California, San Diego, USA.
- 2007-2010. Postdoctoral studies in Computer Science and Engineering, Siberian State Aerospace University, Krasnoyarsk, Russia.
- 2004. Ph.D. in Engineering Sciences (Field of Study– system analysis, management and information processing (computer science, computer facilities and management).
- Krasnoyarsk State Technical University, Krasnoyarsk, Russia.
- 1997. M.S. in Mathematics (Field of Study– Mathematics and computer science).
- Katanov State University of Khakassia, Abakan, Russia.

Research and Professional Experience:

- 2017. The Eighth International Conference on Swarm Intelligence (ICSI'2017) (IEEE Conference Record #41361), Fukuoka, Japan.
- 2015-2017. International Conference on Engineering and Telecommunication (EnT), Moscow, Russia.
- 2016. The Seventh International Conference on Swarm Intelligence (ICSI'2016) (IEEE Conference Record #41361), Bali, Indonesia.
- 2006-2008, 2013-2019. International conference Neuroinformatics, Moscow, Russia.
- 2009. Collaboration with GURU group at Computer Science and Engineering, University of California, San Diego, USA.
- 2007. Object-Oriented Analysis and Design. International Workshop Problems of higher and professional education, Blanes, Spain.

- 2006. Extension courses on information technologies in the Academy of National Economy, Moscow, Russia.
- 2006. Multiparallel Programming. International IEEE EAST-WEST DESIGN & TEST WORKSHOP, Sochi, Russia.
- 2001-2006. All-Russian conference Neuroinformatics and its application, Krasnoyarsk, Russia.
- 2005-2019. Regional conference Problems of Informatization of the Region, Abakan, Russia.
- 1997-2019 Regional conference Katanovskiye Chteniya (that means Reading devoted to the memory of N.F. Katanov), Abakan, Russia.

Professional Appointment:

- Current Appointments 2004-present: Associate Professor of the Department of Information Technologies and Systems, Katanov State University of Khakassia
- Past Appointment 1997-2004: Assistant of the Department of Information Technologies and Systems, Katanov State University of Khakassia

Honors:

1. Research grant "Maximum power tracking system of a photovoltaic module based on the modified fuzzy neuronet" № 19-48-190003 of Russian foundation for basic research and Republic of Khakassia (01/2019-12/2021).
2. Diploma of the Katanov State University of Khakassia, 2018.
3. Research grant "Energy-saving technology of the automated technical system on the basis of the adaptive neurocontroller" №14-41-0402-a of Russian foundation for basic research (01/2014-12/2015).

4. International Fulbright grant «Freeware MegaNeuro», host institution - Computer Science and Engineering, University of California San Diego, USA, 2009 (02/25-08/25).
5. Research grant «Auto summarize system for science's article» of Russian foundation for basic research 2007-2009.
6. Research grant « Intelligent Information Systems for information processing» of Russian foundation for basic research, 2007-2009.
7. Research grant «Intelligent learning machines» of the Katanov State University of Khakassia, 2006-2008.
8. Diploma of the Katanov State University of Khakassia, 2005.
9. Research grant Perfection of Information systems based on artificial neural networks of the Katanov State University of Khakassia, 2003.

Publications from the Last 3 Years:

[1] Engel, E. A. & Engel, N. E. (2018). "Temperature Forecasting Based on the Multi-agent Adaptive Fuzzy Neuronet." *Paper presented at the 18th International Conference Neuroinformatics* 2018, Moscow, Russian Federation, October 8-12.

[2] Engel, E. A. & Engel, N. E. (2019). "Photovoltaic System Control Model on the Basis of a Modified Fuzzy Neural Net" *Paper presented at the 19th International Conference Neuroinformatics* 2019, Moscow, Russian Federation, October 7-11.

[3] Engel, E. A., Kovalev, I. V. & Engel, N. E. (2016). "Control of technical object on the basis of the multi-agent system with neuroevolution and student-teacher of-line learning" *IOP Conference Series: Materials Science and Engineering, 155.* doi:10.1088/1757-899X/155/1/012001. (Engel, Kovalev, 2016).

[4] Engel, E. A., Kovalev, I. V. & Engel, N. E. (2016). "Model of interaction in Smart Grid on the basis of multi-agent system". *IOP Conference Series: Materials Science and Engineering, 155.* doi:10.1088/1757-899X/155/1/012002. (Engel, Kovalev 2016).

[5] Engel, E. A., Kovalev, I. V. & Engel, N. E. (2016). "Intelligent control of PV system on the basis of the fuzzy recurrent neuronet." *IOP Conference Series: Materials Science and Engineering, 122.* doi:10.1088/1757-899X/122/1/012006. (Engel, Kovalev, 2016).

[6] Engel, E. A., Kovalev, I. V. & Engel, N. E. (2016). "Vector control of wind turbine on the basis of the fuzzy selective neural net" *IOP Conference Series: Materials Science and Engineering, 122.* doi:10.1088/1757-899X/122/1/012007. (Engel, Kovalev, 2016).

[7] Engel, E. A. & Kovalev, I. V. (2016). "MPPT of a Partially Shaded Photovoltaic Module by Ant Lion Optimizer". *ICSI 2016, Advances in Swarm Intelligence, Part I*, 9713:382-388. doi:10.1007/978-3-319-41009-8_41. (Engel, Kovalev, 2016).

[8] Engel, E. A. & Kovalev, I. V. (2016). "The Energy Saving Technology of a Photovoltaic System's Control on the Basis of the Fuzzy Selective Neuronet". *ICSI 2016, Advances in Swarm Intelligence, Part II*, 9712:451-457. doi:10.1007/978-3-319-41000-5_45. (Engel, Kovalev, 2016).

[9] Engel, E. A. (2016). "Sizing of a Photovoltaic System with Battery on the Basis of the Multi-Agent Adaptive Fuzzy Neuronet" *Paper presented at International Conference on Engineering and Telecommunication (EnT)*, Moscow, Russia, November 29-30. (Engel, 2016).

[10] Engel, E. A. & Kovalev, I. V. (2017). "Solar Irradiance Forecasting Based on the Multi-agent Adaptive Fuzzy Neuronet." *Advances in Swarm Intelligence. ICSI 2017. Lecture Notes in Computer Science*, 10386:135-140. doi:10.1007/978-3-319-61833-3_14. (Engel, Kovalev, 2017).

[11] Engel, E. A., Kovalev, I. V., Engel, N. E., Brezitskaya, V. V. & Prohorovich, G. A. (2017). "Intelligent control system of autonomous objects". *IOP Conf. Series: Materials Science and Engineering, 173.* doi:10.1088/1757-899X/173/1/012024. (Engel et. al. 2017).

[12] Engel, E. A. (2016). *The intellectual system for forecasting of a non-linear technical object's state.* Certificate about State registration of

computer programs. M.: Federal Service For Intellectual Property (Rospatent). № 2016663468.

[13] Engel, E. A. (2016). *The intellectual controller of a non-linear technical system*. Certificate about State registration of computer programs. M.: Federal Service For Intellectual Property (Rospatent). № 2016663467.

[14] Ekaterina, A Engel., Nikita, E Engel., Alexander, S Degtyarev., Viktor, I Kosenko. & Marina, V Savelyeva. (2018). "Power forecasting for a photovoltaic system based on the multi-agent adaptive fuzzy neuronet". *IOP Conf. Series: Materials Science and Engineering*, 450. doi:10.1088/1757-899X/450/7/072012.
(Engel et. al. 2018).

[15] Engel, E. A. & Engel, N. E. (2018). *The intelligent forecasting of power for a photovoltaic module*. M.: Federal Service for Intellectual Property (Rospatent), Certificate about State registration of computer programs №2018612282.

[16] Ekaterina, A. Engel. (2018). "Modeling, Sizing and Control of a Photovoltaic System on the Basis of a Multi-Agent Adaptive Fuzzy Neuronet" In: *Photovoltaic Systems: Design, Performance and Applications*, edited by Wassila Issaadi, Salim Issaadi, 139-180. Nova Science Publishers. ISBN: 978-1-53613-646-3. Series: Electrical Engineering Developments. (Engel 2018, 139-180).

In: Solar Irradiance
Editor: Daryl M. Welsh

ISBN: 978-1-53618-786-1
© 2020 Nova Science Publishers, Inc.

Chapter 2

DIRECT NORMAL IRRADIANCE: MEASUREMENT, MODELING AND APPLICATIONS

Marius Paulescu[*]
Department of Physics, West University of Timisoara,
Timisoara, Romania

ABSTRACT

The monitoring of solar radiation experienced a vast progress, not only through the expansion of the measurement networks, but also through the improvement of data quality. Nevertheless, the number of stations for monitoring direct normal irradiance (DNI) is still too small for achieving an accurate global coverage. Alternatively, various models for estimating DNI are exploited in many applications. This chapter deals with modeling DNI at the Earth's surface. It is structured in three sections. The first section introduces the methods of measuring DNI at the ground level. The second section summarizes the main classes of models for estimating DNI. The focus shifts upon the parametric class. Within this class, the atmospheric transmittance is explicitly expressed as a function of meteorological parameters. Choosing a highly-performant

---

[*] Corresponding Author's E-mail:marius.paulescu@e-uvt.ro.

model is frequently limited by the availability of the parameters required for its running. Approaches for inferring the parametric models, along with the performances and weaknesses of the current models are reviewed. The third section is devoted to analyzing some applications of DNI. In addition to looking traditionally at DNI as a fuel for concentrating solar systems, its use in computing the sunshine number (a binary indicator stating whether the Sun is shining or not) is discussed.

Keywords: solar irradiance, parametric model, accuracy, sunshine number

INTRODUCTION

The terrestrial atmosphere changes the solar radiation flux during its passing towards the Earth surface through two physical processes: absorption and scattering. As a result, at the ground level, the following physical quantities are measured:

- *Direct-normal irradiance* (DNI) which represents the solar energy flux incoming from the solid angle subtended by the Sun disc on a unitary surface, normal to the Sun direction.
- *Diffuse solar irradiance* (DHI) which represents the solar energy flux incoming from the entire sky, excepting the Sun disc, on a unitary surface that is horizontal.
- *Global horizontal irradiance* (GHI) is the sum of the two components on the horizontal surface:

$$GHI = DNI \cos \theta_z + DHI \tag{1}$$

where θ_z is the zenith angle (i.e., the angle between the Sun direction and the zenith direction at the observation point).

This chapter deals with DNI measurement and modeling. The design and operation of concentrated solar power plants (CSP) are the most quoted applications related to DNI. CSP includes two types of power plants: concentrated photovoltaic (CPV) plants and concentrated solar-thermal plants (CST). A CPV system employs optical devices for increasing the energy flux received on a solar cell surface. A CST plant uses mirrors and/or lenses to guide the solar energy flux into a collecting area, generating heat which is further used to operate an electrical generator. CSPs generate electricity on the basis of DNI, generally DHI having a small to negligible contribution.

The above definition of DNI is generally accepted in atmospheric physics and solar radiative transfer modeling and it is considered further in this chapter. However, there are some differences between the various interpretations of DNI, from atmospheric physics to specific conversion technology. Blanc et al. (2014) emphasize that even if a part of circumsolar radiation (the diffuse radiation coming from the proximity of the Sun disc) is measured by pyrheliometers (see Sec. 2 for pyrheliometer meaning), some concentrating systems use only a small part of it. Therefore, this circumsolar effect must be considered in the solar-resource assessment and CSP performance evaluation. Blanc et al. (2014) conclude that, depending on the topic and scientific field, different understandings of DNI are unavoidable. To allow for a correct interpretation of the published results, it is compulsory to explain clearly which definition of DNI is used.

Usually, DNI is estimated by clear-sky solar irradiance models (see Sec. 3). In meteorology and atmospheric physics, the clear-sky is defined as the sky free of clouds. This definition is generally accepted in the solar radiative transfer field. However, a pyrheliometer measures also DNI in sunny-sky conditions, i.e., no clouds between the Sun and instrument. The connection between DNI under clear-sky and DNI measured under sunny-sky is very complex (e.g., Chauvin et al. 2018), primarily depending on the nature of the measurement instrumentation. This chapter presents the physical basis of measuring DNI followed by an introduction to the measurement instruments (Sec. 2). For gaining further insight into solar irradiance measurement and instrumentation, including those for DNI, we refer the reader to Vignola et al. (2017).

It is important to emphasize that the accurate measurement of DNI is difficult due to the high cost of the performing instruments and their maintenance. Consequently, DNI modeling becomes a key topic for evaluating solar resources in different applications. The estimation of DNI is the main sequence of the algorithms for modeling clear-sky solar irradiance (see e.g., Badescu et al. (2013) for a comprehensive review on clear-sky solar irradiance models). Section 3 of this chapter is devoted to this topic: after a brief introduction to the physics of DNI modeling, three classes of clear-sky solar irradiance models are discussed.

Various facets of DNI modeling were investigated over time. Several examples are given next. Perez-Higueras et al. (2012) noticed the lack of DNI data unlike the availability of GHI data. Subsequently, the authors developed a model for DNI estimation using only GHI as input. Since GHI is the most measured radiometric quantity, such models for converting GHI to DNI are useful tools in practical applications. In a different way, Porfirio and Ceballos (2017) developed a satellite-based model for estimating DNI that uses at input a minimal set of regional meteorological information and avoids empirical adjustment with ground-based radiometric data. The satellite-based model estimates DNI with a spatial resolution of 4 km and it proves good accuracy for climatic and solar resource studies. Looking also to the satellite observations, Eissa et al. (2013) proposed an ANN-based model for the estimation of solar irradiance components, including DNI. The model uses information from the six thermal channels of the SEVIRI instrument (onboard Meteosat Second Generation satellite). Additional inputs are the geographical coordinates and the time stamp. The model shows promising results at a temporal resolution of 15 min and a spatial resolution of 3 km.

DNI MEASUREMENT

Radiometry is the science that studies the measurement of the electromagnetic radiation. An instrument for measuring the flux of electromagnetic radiation is referred to as radiometer. High-quality radiometers of various operation principles and designs were developed for measuring the solar irradiance components (e.g., Vignola et al. 2017). The amazing progress made in recent years in electronics has had a major impact in radiometry.

This section starts with a brief introduction to the principle of measuring DNI. Then, the two common methods of measuring DNI are presented. The section ends with a survey of the online available resources for DNI.

Basis of DNI Measurement

Let us consider a beam of solar rays with a spectral distribution G_λ, normally incident to the surface A of a thermal receiver with mass m, specific heat C and the absorption coefficient α. During a given period dt, the incoming solar energy on the surface A is:

$$\left(\int_0^\infty G_\lambda d\lambda\right) A dt = GA dt \qquad (2)$$

where λ denotes the light wavelength. During the same period dt, the energy absorbed by the thermal receiver is $\alpha GA dt$. A part of this energy is used by the thermal receiver for increasing its temperature, while the other part is dissipated as heat. Let us assume that the temperature of the thermal receiver increases from the ambient temperature T_a to $T = T_a + dT$ while the heat loss by convection is $k(T - T_a) dt$. Even if the heat transfer coefficient k is commonly expressed as a complicated function of A, T and T_a (see e.g., Bergman et al. 2011), for many practical applications it can be approximated as constant, keeping enough accuracy in calculations. The loss by thermal radiation is very small and, in a first instance, it can be neglected. Thus, the thermal equilibrium equation can be written as follows:

$$GA\alpha dt = mCdT + k(T - T_a) dt \qquad (3)$$

Equation (3) can be integrated by using the method of separating the variables, with the following initial condition: at $t = 0$, the temperature of the thermal receiver equals the ambient temperature, i.e., $T(0) = T_a$. After some elementary calculations, the time-evolution equation of the thermal receiver temperature is obtained:

$$T = T_a + \frac{GA\alpha}{k}\left[1 - \exp\left(-\frac{kt}{mC}\right)\right] \quad (4)$$

If the solar energy flux is constant, after a long enough time the system will reach a steady state, characterized by the equilibrium temperature:

$$T_m = T_a + \frac{GA\alpha}{k} \quad (5)$$

One of the most important requirements for a radiometer is related to its response time, which must be as short as possible. As a result, the time it takes for the system to reaches the steady state, $\tau = mC/k$, must be as small as possible. It can be obtained only by reducing the product mC, because an increasing of k leads in Eq. (5) to a decreasing of the temperature amplitude $T_m - T_a$, this affecting the measurement precision.

Let us assume now that after the system reached the thermal equilibrium, the optical flux is suppressed. The thermal equilibrium equation (Eq. 3) becomes:

$$0 = mCdT + k(T - T_a)dt \quad (6)$$

with the solution $T = T_a + \frac{GA\alpha}{k}\exp\left(-\frac{kt}{mC}\right)$, which expresses the system reversibility: at a time $t \gg \tau$ the system returns back to the initial state $T = T_a$. The merit of this proofs is Eq. (5) which shows a linear dependence between the temperature of the thermal receiver and the optical flux density G incident on its surface. Therefore, the difference $T - T_a$ represents a measure of solar energy flux. Eq. (5) represents the basic equation of the solar irradiance measurement.

Methods of Measuring DNI

Two methods for measuring DNI and the associated instruments are presented next.

The Pyrheliometer Method

This is the most direct method for measuring DNI, the pyrheliometer being a broadband radiometer employed for such measurements.

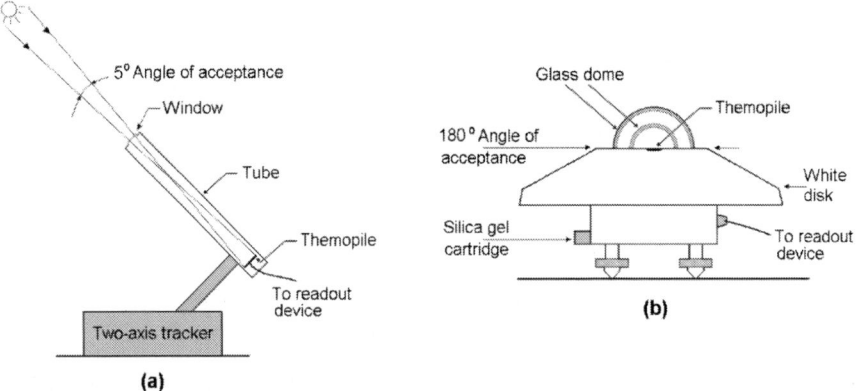

Figure 1. Schematic diagram of a (a) pyrheliometer and (b) pyranometer.

Figure 1a shows the diagram of a pyrheliometer. The sensor consists of a thermal receiver (typically a multi-junction thermopile) placed to the bottom of a collimating tube which is equipped with a quartz window to protect the sensor. The sensor is coated with optical black paint (acting as a perfect absorber in the wavelengths range 0.28 to 3 μm). The temperature of the sensor is compensated to minimize sensitivity to the ambient temperature variations. The pyrheliometer aperture angle is 5°. Consequently, the solar radiation flux is received only from the Sun disc and a very limited circumsolar region. The diffuse radiation originated from the rest of the sky is excluded. The instrument should be permanently pointed towards the Sun. A two-axis Sun tracking mechanism is very often used for this purpose. A readout device is used with the pyrheliometer sensor to display/record the actual value of DNI.

The Pyranometric Method

This is an indirect method. GHI and DHI are measured directly with pyranometers, then DNI being computed on the basis of Eq. (1). A pyranometer is a broadband radiometer primarily designed for measuring GHI. Figure 1b shows a diagram of a typical pyranometer. It consists of a thermal receiver shielded against weather threats (rain, wind and dust) by two concentric hemispherical transparent covers made of glass. A white disc limits the acceptance angle to 180°. A small cartridge of silica gel inside the inner-dome absorbs water vapor. By removing the contribution of DNI, a pyranometer can also be used for measuring DHI. For this, a small shading disc is mounted on a solar tracker, ensuring a permanent shading of the sensor. Alternatively, a shadow ring may prevent whole day long the direct radiation to reach the sensor. Because the maximum of the Sun elevation angle changes day by day, it is necessary to adjust periodically (days lag) the height of the shadow ring. On the other hand, because the shadow ring intercepts a part of the diffuse radiation, it is necessary to correct the measured values. The percentage of DHI intercepted by the shadow ring varies during a year with its position and atmospheric conditions (Siren 1987). Kotti et al. (2014) evaluated the performance of four procedure for adjusting DHI measured by a pyranometer equipped with a shadow ring. The approach consists in comparing the DNI values measured simultaneously using the pyranometric and pyrheliometric methods. The authors conclude that the empirical procedure developed by Batlles et al. (1995) performs best.

Figure 2. SMP10 pyranometers measuring global solar irradiance (left) and diffuse solar irradiance (right) on the Solar Platform of the West University of Timisoara, Romania.

Figure 2 shows two SMP10 pyranometers (https://www.kippzonen.com/), operating on the Solar Platform of the West University of Timisoara, Romania (Solar Platform 2019). According to ISO9060, the SMP10 pyranometer is classified as secondary standard. To maximize the measurements quality, the pyranometer is mounted in a heated and ventilated box. The pyranometer from the left side measures GHI while the pyranometer from the right side, equipped with a shadow ring, measures DHI. The pyranometer base is mounted parallel to the ground while the shadow-ring is positioned in such a way that its axis is parallel to the Earth's axis.

The accepted classification of the pyranometers with respect to their quality is defined by the International Standard ISO 9060/1990 that is also adopted by the World Meteorological Organization (WMO 2008). ISO 9060 standard distinguishes three classes of pyranometers: the best is called (somewhat improperly) *secondary standard*, the second best is called *first class* and the third one - *second class*. Well-maintained secondary standard pyranometers are used as reference pyranometers for field-calibration of the pyranometers from first and second class.

The quality control of solar irradiance data consists routinely of three stages: (1) *Quality assurance*. This stage includes measures as the suitable selection and installation of an instrument, regular calibration and maintenance. The first stage is very important since once a datum is acquired, it remains forever as is. Chapters 7 and 8 of the WMO Guide to Meteorological Instruments and Methods of Observation (WMO 2008) provide guidance for carrying out accurate solar radiation measurements. (2) *Real-time assessment of data during the measurement process*. Such tests indicate whether a data value is missing, its value is reasonable, too small or too large. In addition to testing physical limits, a quality control criterion can be applied. For instance, GHI must equal the sum of diffuse and beam irradiances (Eq. 1). (3) *Retrospective data analysis*. The test is performed by testing the measured data against reliable physical models. A review of the procedures for quality assessment of solar irradiation data was reported by Younes (2005).

The performance of 51 commercially available and prototype radiometers used for measuring GHI or DNI is assessed by Habte et al. (2015). The radiometers were deployed for one year and their measurements were compared under clear sky, partly cloudy, and mostly cloudy conditions to reference values. Under clear-sky conditions and $\theta_z < 60°$, differences of less than ±5% were observed among all GHI and DNI measurements when compared to the reference radiometers.

The most important source of DNI data with global coverage and online access is the Baseline Surface Radiation Network – BSRN (https://bsrn.awi.de/). BSRN is operating by the World Radiation Monitoring Center (WRMC). WRMC is run by the Division of Climate Sciences at the Alfred Wegener Institute, Helmholtz Centre for Polar and Marine Research in Bremerhaven, Germany. BSRN started operating in 1992 with 9 stations. Currently (July 2019), the network comprises 52 stations in operation covering the whole Earth, 12 stations temporarily or permanently closed and 10 stations with a candidate status. BSRN collect and centrally archive high-quality ground-based solar irradiance measurements (at least DNI, GHI and DHI) at 1 min resolution. The BSRN data can be accessed online at https://bsrn.awi.de/data/data-retrieval-via-pangaea/. A list of publications concerning the development of BSRN as well as a collection of papers reporting studies based on BSRN data are available at https://bsrn.awi.de/other/publications/.

DNI MODELING

This section deals with the estimation of DNI. Three classes of models are discussed: spectral codes, parametric models and empirical equations. The focus shifts upon the parametric class. Within the parametric class, the atmospheric transmittance is considered as an average value over the entire solar spectrum. It is explicitly expressed as a function of the surface meteorological parameters and/or the atmospheric column content. Approaches for inferring the parametric models, their strengths and weaknesses as well as the input's influence on the estimates are discussed. The section ends with a comparative study on the accuracy of estimates provided by different models.

Extraterrestrial Radiation

Since the Earth moves around the Sun on an elliptical trajectory, the density of the solar radiation flux at the upper limit of the atmosphere varies continuously over time. At the mean Sun-Earth distance (1 AU) the density of the solar energy flux at the upper limit of the atmosphere is commonly referred to as "solar constant" G_{SC}. Based on a 42-year total solar irradiance time series and a reconciliation of spaceborne observations the most recent evaluation of the solar constant is $G_{SC} = 1361.1$ W/m² (Gueymard 2018). It is with 5 W/m² lower than the previous accepted value 1366.1 W/m², dating back from the early 2000s. At an arbitrary distance between the Sun and Earth, the density of the solar energy flux at the upper limit of the atmosphere (extraterrestrial irradiance) is routinely expressed as the solar constant multiplied by a correction factor:

$$G_{n,ext} = G_{SC}\varepsilon \tag{7}$$

The correction factor can be assumed to be constant over one day for the most solar applications. It is well approximated by the equation

$\varepsilon = 1 + 0.0342 \cdot \cos\left[\frac{2\pi}{365}(j-1)\right]$, where the index $j = 1...365$ stands for the Julian day (i.e., the day number within the year). The extraterrestrial solar irradiance on a horizontal surface is computed simply using the cosine law:

$$G_{ext} = G_{n,ext} \cos\theta_z \qquad (8)$$

where the zenith angle θ_z is expressed with respect to the geographical latitude ϕ, the Sun's declination angle δ and the hour angle ω (see e.g., Ch. 5 from the book Paulescu et al. (2013) for derivation):

$$\cos\theta_z = \sin\phi\sin\delta + \cos\phi\cos\delta\cos\omega \qquad (9)$$

Since the declination angle changes less than 0.5° per day, it can be also assumed to be constant over one day. For a given Julian day j it can be calculated with the equation $\delta[\text{deg}] = 23.45\sin\left[\frac{2\pi}{365}(j+283)\right]$.

The hour angle ω stands for the angular displacement of the Sun to East or West in respect to the local meridian. By convention ω takes negative values in the morning and positive in the afternoon. Mathematically, it is expressed as:

$$\omega = \frac{2\pi}{24}(t-12) \qquad (10)$$

where t (in hours) is the solar time or the solar true time.

Solar Radiation through the Atmosphere

During its passing through the atmosphere towards the Earth's surface, the solar radiation flux is changed by two physical processes: scattering and absorption. The atmospheric gases and aerosols scatter a part of the incoming solar radiation in random directions without any alteration of the wavelength. At first instance, the amount of scattered radiation depends on two factors: the wavelength of the incoming radiation and the size of the scattering particles. A non-negligible fraction of the scattered radiation is redirected back to space. The atmospheric gases also absorb a part of the incoming solar radiation. While scattering is a continuous phenomenon with respect to wavelength, the atmospheric gases absorb selectively. Thus, the absorption effects are very complex in the whole solar spectrum and the process is harder to be treated analytically than the scattering process.

Modeling the Effects of Cloudless Atmosphere on Solar Radiation

Firstly, let us to evaluate the attenuation of the solar energy flux at wavelength λ by an atmospheric layer of thickness dz, located at an altitude z. With the notations from Fig 3, the variation of the solar radiation energy flux density between the planes z and $z + dz$ at wavelength λ can be written on basis of the Beer-Lambert law $dG_\lambda = t_\lambda(z) G_\lambda dz$, where $t_\lambda(z)$ is the monochromatic extinction coefficient.

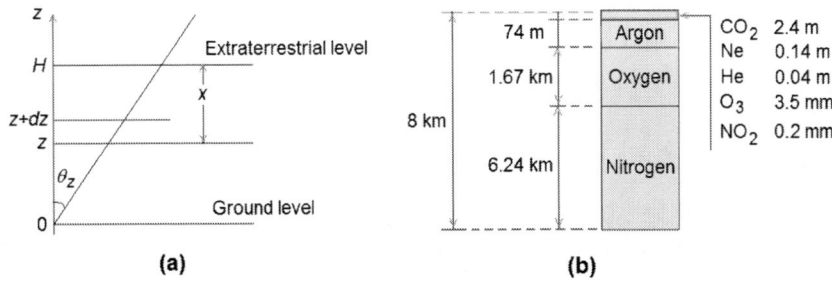

Figure 3. (a) Schematic to the calculation of the solar radiation flux attenuation by an atmospheric layer of thickness x. (b) The thickness of the various gas layers in the standard atmosphere.

The solar radiation energy fluxes at the altitudes z_1 and z_2 are related by the equation:

$$G_\lambda(z_2) = G_\lambda(z_1)\exp\left(\int_{z_1}^{z_2} t_\lambda(z)dz\right) \tag{11}$$

which is obtaining by integrating the Beer-Lambert equation between z_1 and z_2. The extinction coefficient $t_\lambda(z)$ encapsulates the effects of both processes, absorption and scattering. Given that these processes are independent, $t_\lambda(z)$ can be split into two distinct terms $t_\lambda(z) = t_a(\lambda,z) + t_d(\lambda,z)$, where $t_a(\lambda,z)$ models the absorption processes and $t_d(\lambda,z)$ models the scattering processes.

In normal atmospheric conditions, at low pressure and high temperature, the atmospheric gases behave qualitatively like an ideal gas. Thus, an atmospheric gas (indexed by the subscript i) is characterized by the absorption coefficient $a_i(\lambda)$ and the diffusion coefficient $D_i(\lambda)$, which depend only on the gas nature. The contribution to the extinction coefficient of the gas i is proportional with its concentration $n_i(z)$ at altitude z. Under these assumptions we can writte:

$$t_a(\lambda,z) = \sum_i n_i(z)a_i(\lambda), \quad t_d(\lambda,z) = \sum_i n_i(z)D_i(\lambda) \tag{12a,b}$$

The Earth's atmosphere is not uniform, its density, pressure and temperature varying with altitude. For modeling rationale, it is suitable to replace the real atmosphere with an effective atmosphere which assume $n_i(z)$ a constant for each gas i. An appropriate model is the standard atmosphere which consists of a homogenous gas layer with the same composition as the real atmosphere, but all the gases are at normal pressure. The vertical height of the standard atmosphere is $H \approx 8$ km. Figure 3b shows the thickness of gases layers in the standard atmosphere, if they are separated. In the standard atmosphere the gas i is homogenous distributed with the concentration $\bar{n}_i$ and the coefficients $a_i(\lambda)$ and $D_i(\lambda)$ depend only on the wavelength. Thus, for a vertical crossing of the solar radiation flux through atmosphere (see Figure 3a), the integral in Eq. (11) can be evaluated as follows:

$$\int_H^z \sum_i n_i(z)\left[a_i(\lambda)+D_i(\lambda)\right]dz = (z-H)\sum_i \bar{n}_i\left[a_i(\lambda)+D_i(\lambda)\right]$$

Replacing the above result in Eq. (11) and taking $z_1 = H$ and $z_2 = z$, it becomes:

$$G_\lambda(z) = G_\lambda(H)\exp\left[-(H-z)\left(\sum_i \bar{n}_i a_i(\lambda)+\sum_i \bar{n}_i D_i(\lambda)\right)\right] \quad (13)$$

Denoting the path length of solar radiation through the atmosphere $x = H - z$ and $T_a(\lambda)+T_d(\lambda) = H\left(\sum_i \bar{n}_i a_i(\lambda)+\sum_i \bar{n}_i D_i(\lambda)\right)$, Eq. (13) becomes:

$$G_\lambda(z) = G_{ext}(\lambda)\exp\left[-\frac{x}{H}\left(T_a(\lambda)+T_d(\lambda)\right)\right] \quad (14)$$

Equation (14) models the effect of the Earth's atmosphere on the solar radiation flux passing the atmosphere on the zenith direction ($\theta_z = 0$). At a zenith angle different from zero, the path length x is visibly longer. The ratio $m = x/H$ defines the atmospheric optical mass and, at first instance ($\theta_z < 85°$) m is approximated well by $1/\cos\theta_z$ (see Figure 3a). A more accurate equation, which correct this approximation near to sunrise and sunset and frequently used in practice, is (Kasten and Young 1989):

$$m = \frac{1}{\cos\theta_z + 0.50572(96.07995-\theta_z)^{-1.6364}} \quad (15)$$

The quantities:

$$\tau_a(\lambda) = \exp\left[-mT_a(\lambda)\right] \text{ and } \tau_d(\lambda) = \exp\left[-mT_d(\lambda)\right] \quad (16a,b)$$

are called spectral atmospheric transmittances. As defined, $\tau_a(\lambda)$ and $\tau_d(\lambda)$ encapsulate the effects of the absorption and scattering, respectively, being the fundamental physical quantities in modeling solar radiation at the Earth surface. Different models for DNI are differentiated just by the way in which they themselves express $\tau_a(\lambda)$ and $\tau_d(\lambda)$. Three different classes of models for DNI, spectral, parametric and empiric, are discussed next.

Spectral Models

Considering Eq. (14), a specific atmospheric transmittance can be associated to every extinction process caused by a certain species of particles from the atmosphere. Commonly, the spectral models consider two independent scattering processes (Rayleigh and due to aerosols) and three independent absorption processes (ozone, water vapor and all other atmospheric constituents, generically called mixed gas). Rayleigh scattering occurs when solar radiation interacts with particles with dimensions much smaller than its wavelength. The particles may be individual atoms or molecules. Aerosols are constituted by small particles with dimensions ranging between 0.02 and 10 µm. In general aerosols scatter the solar radiation; only a small amount being absorbed.

Leckner's spectral model (Leckner 1979) can be considered a milestone in the history of spectral solar irradiance models developed for computerized engineering calculations. Many other models reported in literature are based on the Leckner's model equations (e.g., Yang et al. 2001; Paulescu and Schlett 2003). The Leckner's model considers five independent processes experienced by solar radiation passing through the atmosphere: Rayleigh scattering $\tau_R(\lambda)$, aerosols scattering $\tau_a(\lambda)$, water vapour absorption $\tau_w(\lambda)$, ozone absorption $\tau_{O3}(\lambda)$ and mixed gases absorption $\tau_g(\lambda)$. The mathematical equations of the transmittances are:

$$\tau_R(\lambda) = \exp\left[-0.008735m\frac{p}{p_0}(\lambda)^{-4.08}\right]; \; \tau_a = \exp\left[-m\beta\lambda^{-1.3}\right] \quad (17a,b)$$

$$\tau_w(\lambda) = \exp\left[-0.2385mw\left(1+20.07mwK_w(\lambda)\right)^{-0.45} K_w(\lambda)\right] \quad (18a)$$

$$\tau_{O3}(\lambda) = \exp\left[-ml_{O3}K_{O3}(\lambda)\right] \quad (18b)$$

$$\tau_g(\lambda) = \exp\left[-1.41m\left(1+118.3mK_g(\lambda)\right)^{-0.45} K_g(\lambda)\right] \quad (18c)$$

where m is the atmospheric optical mass, l_{O3} is the ozone column content, β is the Ångström turbidity coefficient, p is the site-specific atmospheric pressure and w [g/cm^2] is the water vapor column content. The extinction coefficients $K_{O3}(\lambda)$, $K_w(\lambda)$ and $K_g(\lambda)$ are tabulated (Leckner 1978).

The direct-normal solar irradiance at a wavelength λ is naturally expressed (in the sense of Eq. 14) as follows:

$$G_n(\lambda) = G_{n,ext}(\lambda) \cdot \tau_{O3}(\lambda)\tau_w(\lambda)\tau_g(\lambda)\tau_R(\lambda)\tau_a(\lambda) \quad (19)$$

In addition to geographical coordinates and temporal information, the estimation of DNI spectrum requires a set of three surface atmospheric parameters: the ozone column content, the Ångström turbidity coefficient and the water vapor column content. The ozone column content can be retrieved from the NASA Ozone Watch website https://ozonewatch.gsfc.nasa.gov/ while the Ångström turbidity coefficient can be retrieved from the Aerosol Robotic Network website https://aeronet.gsfc.nasa.gov/.

Figure 4 shows the Leckner's atmospheric transmittances with respect to wavelength. The graphs emphasizes the weight of different extinction processes on the spectral solar radiation flux. The scattering processes are continuous with respect to wavelength, mainly the shortwave radiation being scattered. Most ultraviolet radiation is absorbed by ozone; for wavelengths below 280 nm the absorption being complete. While in the visible solar spectrum the absorption is relatively low, it is strong in the infrared, mainly due to vibrations and rotations of water vapour molecules. Figure 5 illustrates the spectral distribution of DNI at ground level. The weights of absorption and scattering processes are emphasized.

DNI is simply computed by integrating Eq. (19) over the entire spectrum:

$$DNI = \int_0^\infty G_n(\lambda) d\lambda \qquad (20)$$

Generally, the absorption coefficients are tabulated, and the integral is computed as a Darboux sum. The usual limits of integration are $\lambda_{min} = 0.28\,\mu m$ and $\lambda_{max} = 4\,\mu m$. DNI estimated by the Leckner's model is illustrated in Figure 6. As the atmospheric air mass increases, i.e., the pathlength of solar radiation increases, DNI decreases, as expected. Depending on the values of the atmospheric parameters the decreasing can be slow or fast. The Ångström turbidity coefficient, quantifying the atmospheric aerosol loading, is the most influential parameter.

At an atmospheric air mass greater than 2, a high atmospheric aerosol loading ($\beta = 0.2$) may reduce to half the estimated value of DNI in the same atmosphere but roughly without aerosols ($\beta = 0.01$) (Figure 6a). Compared to aerosols, the water vapour column content has a lower impact on DNI. Figure 6b shows a decreasing of DNI with approximative 200W/m² when the water vapour column content increases from $w = 0.5\,g/cm^2$ to $w = 4\,g/cm^2$. The ozone column content have a small impact in DNI (Figure 6c).

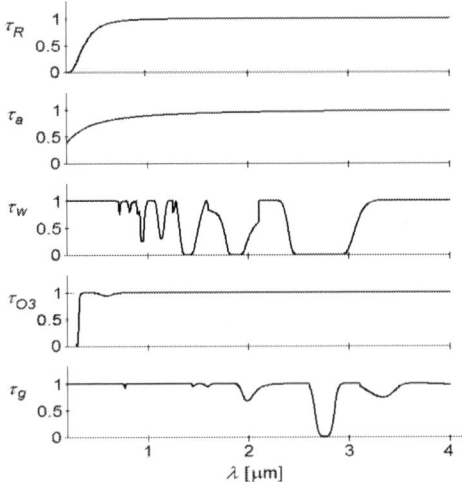

Figure 4. Spectral atmospheric transmittances in the Leckner's model with respect to wavelength λ: Rayleigh scattering τ_R, aerosols scattering τ_a, water vapour absorption τ_w, ozone absorption τ_{O3} and mixed gases absorption τ_g. The graphs are built with the following atmospheric parameters: $m = 1.5$, $\beta = 0.079$, $w = 1.3 \text{ g/cm}^2$ and $l_{O3} = 0.35 \text{ mm}$.

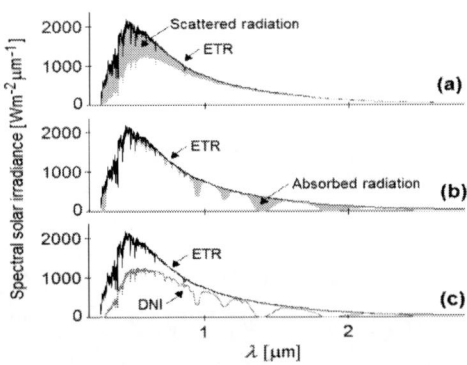

Figure 5. Spectral distribution of solar irradiance at extraterrestrial level (ETR) and (a) the scattered energy (in grey); (b) the absorbed energy (in grey) and (c) the spectral distribution of the direct-normal irradiance DNI at ground level. The graphs are built with the following atmospheric parameters: $m = 1.5$, $\beta = 0.079$, $w = 1.3 \text{ g/cm}^2$ and $l_{O3} = 0.35 \text{ mm}$.

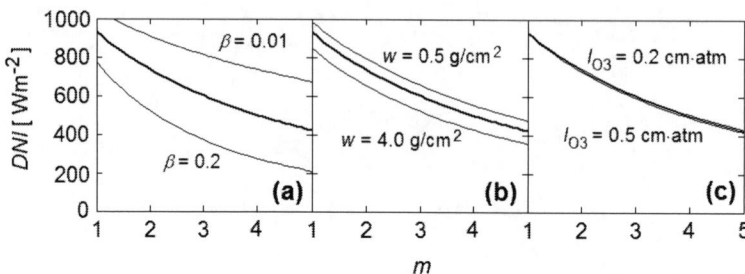

Figure 6. Direct-normal irradiance DNI with respect to the atmospheric air mass m. The emphasized curve corresponds to the following atmospheric parameters: $m = 1.5$, $\beta = 0.079$, $w = 1.3\ g/cm^2$ and $l_{O3} = 0.35\ mm$. The same parameters were used to build the boundary curves, excepting one parameter: (a) the Ångström turbidity coefficient β; (b) the water vapour column content w and (c) the ozone column content l_{O3}.

It is worth to mention here the study by Nikitidou et al. (2014) which evaluates the effect of spatial and temporal variability of aerosol optical depth (AOD) on DNI under clear sky. The study, focused over Europe, shows that the aerosol effect on DNI is higher in areas experienced desert dust intrusions and intense anthropogenic activities, such as the Mediterranean basin and the Po Valley in Italy. Over these areas the reduction of DNI due to aerosols can reach values of 45%. Due to the change in AOD, the value of daily DNI can be with 20% smaller than the corresponding monthly climatological value. The study by Nikitidou et al. (2014) highlights the need for reliable measurements of AOD and a correct representation of aerosols in modeling DNI.

We detailed here the Leckner model both due to its relevance and simplicity. More sophisticated and accurate models were developed over the last 40 years. Here we mention only the spectral solar irradiance model SMARTS (Gueymard 1995) which in 2019 celebrated 25 years (Gueymard 2019). Compared to Lecner's model the extraterrestrial spectrum was improved, both in accuracy and resolution, containing a total of 1881 measurements at 1 nm wavelength intervals between 0.28 μm and 1.7 μm and at 5 nm intervals between 1.705 μm and 4 μm. Only 70 wavelengths were used in the Leckner's work. SMARTS2 includes high accurate absorption coefficients and it introduces more accurate transmittance functions for all the atmospheric extinction processes, considering the temperature and humidity effects. The nitrogen dioxide (NO₂) is added to the list of the independent atmospheric absorbers. Gueymard (2019) surveys the diverse applications in which the SMARTS model has been involved during the last 25 years, providing an overview of the model's influential status over a broad range of scientific disciplines.

Parametric Models

Spectral solar irradiance models provide high accuracy in estimating DNI. Such models are widely used in atmospheric and climatic research. Simplified models are preferred in engineering applications. The parametric solar irradiance models form a class of such simple models. These models are mainly inferred from spectral solar irradiance codes by averaging the spectral atmospheric transmittances with respect to wavelength. A parametric model estimates solar irradiance with an equation formally similar to Eq. (14), but with the dependence on wavelength removed. In principle, DNI is expressed as follows:

$$G_n = G_{SC} \varepsilon \prod_i \bar{\tau}_i \qquad (21)$$

The key terms in Eq. (21) are the average atmospheric transmittances $\bar{\tau}_i$ associated to the independent extinctions processes considered by the model. The subscript i counts the process. $\bar{\tau}_i$ are usually computed by a weighted average of the corresponding spectral transmittance:

$$\bar{\tau}_i = \int_0^\infty \tau_i(\lambda) G_{ext}(\lambda) d\lambda \bigg/ \int_0^\infty G_{ext}(\lambda) d\lambda \qquad (22)$$

For broadband solar irradiance models, the integration is performed typically between 0.28 μm and 4 μm. As illustration, the PS model is presented next.

The PS model (Paulescu and Schlett 2003) was built by integration the Leckner's spectral solar irradiance model. All the five transmittances are calculated with the equation:

$$\bar{\tau}_i = \exp\left[-x_i\left(a_i + b_i x + c_i x_i^{-d_i}\right)\right] \quad (23)$$

where the subscript i stands for R (Rayleigh scattering), a (aerosol scattering), O3 (ozone absorption), w (water vapor absorption) and g (mixed gases absorption). The coefficients a_i, b_i, c_i, d_i and the corresponding variables x_i are listed in Table 1. Thus, the PS model estimates DNI with the equation:

$$G_{n,PS} = G_{SC}\varepsilon \cdot \bar{\tau}_{O3}\bar{\tau}_w\bar{\tau}_g\bar{\tau}_R\bar{\tau}_a \quad (24)$$

Table 1. The variable x_i and the coefficients a_i, b_i, c_i, d_i from Eq. (23)

i	x	a	b	c	d
R	m	0.709	0.0013	-0.5856	0.058
a	$m\beta$	1.053	-0.083	0.3345	-0.668
O3	ml_{O3}	0.0184	-0.0004	0.022	-0.66
w	$m \cdot w$	-0.002	$1.67 \cdot 10^{-5}$	0.094	-0.693
g	m	$-5.41 \cdot 10^{-5}$	$-3.8 \cdot 10^{-6}$	0.0099	-0.62

The results reported by Paulescu and Schlett (2003) demonstrate that the PS model performs with reasonable accuracy (overall tests show nRMSE < 10%). Badescu et al. (2013) tested 54 solar irradiance models and concluded that the PS model is one of the best broadband models.

Empirical Models

Many clear-sky solar irradiance models are constructed empirically by fitting the radiometric data recorded in a given geographical location. This entails a high local specificity of these models. However, various tests show that an empirical model can be applied in a location with similar climate as the origin place. Generally, the empirical models require at input only the geographical coordinates and the time stamp without any measured weather parameter. Being very simple, these models are widely preferred in various applications. As illustration, two empirical models for DNI are presented next.

The Hottel empirical model (Hottel 1976) estimates DNI with reasonable accuracy using at input only altitude. The model equation is:

$$G_{n,H} = G_{SC} \varepsilon \left[a_1 + a_2 \exp\left(-\frac{a_3}{\cos \theta_z}\right) \right] \qquad (25)$$

where the fitted coefficients are expressed as functions of the altitude z (in km): $a_1 = 0.4327 - 0.00821(6-z)^2$, $a_2 = 0.5055 + 0.00595(6.5-z)^2$ and $a_3 = 0.2711 + 0.01858(2.5-z)^2$. The Hottel model is applicable up to 2.5 km altitude.

The Biga and Rosa empirical model (Biga and Rosa 1979) estimates DNI on the basis of only geographical coordinates and the time:

$$G_{n,BR} = 926.0 (\sin h)^{0.29} \qquad (26)$$

DNI Models Performance

In the last decades a permanent focus was maintained on increasing the performance of DNI models. This is well captured by some review papers dealing with the comparative analysis of DNI models performance. For example, Gueymard and Ruiz-Arias (2015) assess the performance of 24 models used to estimate DNI under clear-sky at a 1-min time resolution in arid environment. The results show that the performance of the parametric models decreases when the atmospheric turbidity pass beyond the typical conditions in temperate climate. The study shows that, assuming an ideal model, DNI can be estimated within a 5% accuracy, only if the Ångström turbidity coefficient is known to within 0.02. Benkaciali et al. (2018) compare 18 clear-sky solar irradiance models for estimating DNI. The models were tested against data measured at two stations in Algeria in various weather conditions. The first conclusion is that the turbidity parameters (e.g., the aerosol optical depth, the Ângström turbidity coefficient) have the highest influence on the models accuracy. The authors hierarchize the models on basis of several statistical indicators of accuracy aiming for an appropriate selection of a DNI model in solar power applications.

In order to provide the reader with a concrete image of the clear-sky solar irradiance models accuracy applied to DNI estimation, an example is presented next. The four models discussed in this this section, Leckner (Eq. 20), PS (Eq. 24), Hottel (Eq. 25), Biga and Rosa (Eq. 26), were tested against of a challenging set of measured data (Table 2). Radiometric data were retrieved from BSRN while the inputs for the Leckner and PS models were retrieved from AERONET. The single condition at the selection of these stations was that a BSRN station to be located in the proximity of an AERONET station. Table 2 lists the stations, their geographical coordinates, local climate (arid or temperate, according to the Köppen-Geiger climate classification) and the number of the measurements from every station. Since the four models (Leckner, PS, Hottel, Biga and Rosa) are basically clear sky-solar irradiance models, the tests were performed against data measured under clear sky. The procedure of clear-sky data selection is described by Calinoiu et al. (2018). Due to the different temporal resolution of the two networks, BSRN and AERONET, a further processing of data series was required. While in BSRN solar irradiance is measured at equal time intervals of one minute, the parameters on the AERONET are measured at equal time intervals of 15 minutes. Thus, after synchronizing the two sets of data, at a given station, in a given day only 50 - 60 instantaneous records were retained, DNI data being located on a nice and smooth bell shape.

The models performance was assessed in terms of two statistical indicators very often used in solar radiation modeling: normalized root mean square error *nRMSE* and normalized mean bias error *nMBE*:

$$nRMSE = \left[M \sum_{i=1}^{M} (c_i - m_i)^2 \right]^{1/2} \bigg/ \sum_{i=1}^{M} m_i \qquad (27a)$$

$$nMBE = \sum_{i=1}^{M} (c_i - m_i) \bigg/ \sum_{i=1}^{M} m_i \qquad (27b)$$

where c and m refer to computed and measured values, respectively, while M is the sample size.

Table 2. Data used for testing the DNI models. N represents the number of measurements collected from each station while KG indicate main climate class (B arid and C temperate) according to Köppen-Geiger classification (Kottek et al. 2006)

Location	KG	N	BSRN station			AERONET station		
			Lat. [deg]	Long. [deg]	Alt. [m]	Lat. [deg]	Long. [deg]	Alt. [m]
Tamanrasset/Algeria	B	548	22.79	5.52	1385	22.79	5.53	1377
Solar Village/Saudi Arabia	B	424	24.91	46.41	650	24.90	46.39	764
Tucson / USA	B	370	32.22	-110.95	786	32.23	-110.95	779
Petrolina/ Brasil	B	287	-9.06	-40.31	387	-9.38	-40.50	370
Sede Boqer/ Israel	B	548	30.85	34.77	500	30.85	34.78	480
Boulder/ USA	C	394	40.05	-105.00	1577	40.05	-105.00	1604
Palaiseau/ France	C	584	48.71	2.20	156	48.70	2.20	156
Carpentras/ France	C	368	44.08	5.05	100	44.08	5.05	100

Table 3. Performance of the clear-sky models Leckner (Eq. 20), PS (Eq. 24), Hottel (Eq. 25), Biga and Rosa (Eq. 26), applied to the DNI estimation at the stations from Table 2

Data	nRMSE				nMBE			
	L	PS	H	BR	L	PS	H	BR
All	0.095	0.1293	0.179	0.192	-0.002	-0.089	0.011	0.037
Climate B	0.113	0.165	0.208	0.247	-0.044	-0.129	0.085	0.078
Climate C	0.072	0.076	0.143	0.109	0.044	-0.045	-0.070	-0.008
Solar Village	0.155	0.216	0.312	0.362	-0.144	-0.204	0.272	0.303
Palaiseau	0.069	0.078	0.167	0.082	0.047	-0.056	-0.153	-0.031

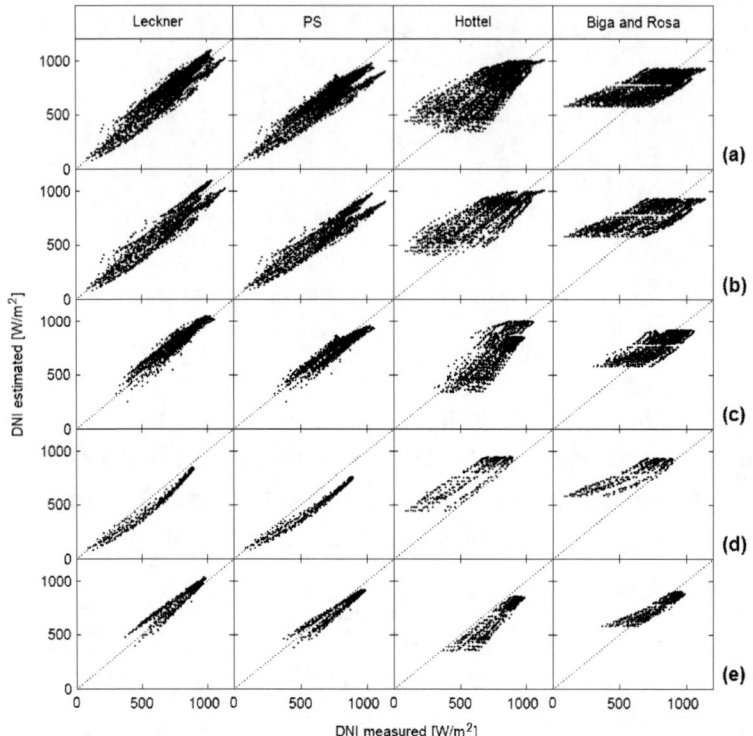

Figure 7. Direct-normal irradiance DNI estimated by the Leckner (L), Paulescu and Schlett (PS), Hottel (H) and Biga and Rosa (BR) models against five sets of measured data (Table 2): (a) all data, (b) data collected in arid climate, (c) data collected in temperate climate, (d) data measured at Palaiseau and (e) data measured at Solar Village.

Table 3 and Figure 7 summarizes the results of testing the models. As expected, the spectral and parametric models perform better than the empirical models. The inputs ascribe to the parametric models more flexibility in capturing the atmosphere influence on solar irradiance. The atmospheric aerosol loading strongly influence the models' accuracy. For example, the Leckner's model achieved $nRMSE$ = 0.113 in arid climate (the average value of the Ångström turbidity coefficient over the dataset is $\bar{\beta}$ = 0.150) while $nRMSE$ = 0.072 in temperate climate ($\bar{\beta}$ = 0.042). Looking at individual stations the differences are more visible: at Solar Village ($\bar{\beta}$ = 0.268) $nRMSE$ = 0.264 while at Palaiseau ($\bar{\beta}$ = 0.037)

$nRMSE = 0.069$. The general conclusion is that there is room enough for enhancing the aerosol representation in clear-sky solar irradiance models.

APPLICATIONS

DNI is the relevant component of solar irradiance for designing and operating the concentrating solar plants. Instead of looking at DNI traditionally, as a fuel for concentrating solar systems, in this section we will look at the role of DNI in defining what "the Sun shines in the sky" means. So far, we have been concerned with the estimation of DNI. From a different perspective, a brief discussion on forecasting DNI is inserted in this section.

Sunshine Number

Two simple measures are often used for quantifying the state of the sky during a given period. The straightforward quantifier is the total cloud cover amount, which represents the fraction of the celestial vault covered by clouds. The second simple quantifier is the relative sunshine. Let's denotes by s the sunshine duration during a given time interval S. The relative sunshine is defined as $\sigma = s/S$. Obviously, a low value of σ is an indication for a high cloud cover amount. A simple parameter describing the instantaneous state of the sky, namely the sunshine number SSN, has been defined by Badescu and Paulescu (2011a). It is defined as a time dependent random binary variable, as follows:

$$SSN_t = \begin{cases} 0 & \text{if the sun is covered by clouds at time t} \\ 1 & \text{otherwise} \end{cases} \quad (28)$$

Badescu and Paulescu (2011a) treats the statistical properties of SSN while Badescu and Paulescu (2011b) presents the elementary statistical and sequential properties of SSN. The average value of SSN over a given period Δt equals the relative sunshine σ during Δt, i.e., $\overline{SSN} = \sigma$.

SSN measurement is based on DNI. Series of SSN values are derived from the series of DNI values using the WMO sunshine criterion (WMO 2008): The Sun shines at time t if DNI exceeds 120 W/m²:

$$SSN_t = \begin{cases} 1 & \text{if } DNI_t > 120 \text{ W/m}^2 \\ 0 & \text{otherwise} \end{cases} \quad (29)$$

Thus, the WMO sunshine criterion transforms the DNI series (of real numbers) in a binary series. Two applications of SSN series are presented next.

Quantification of the Solar Irradiance Variability

A quantifier for the variability in the solar irradiance series based on the number of changes that SSN exhibits during a time interval Δt has been introduced by Paulescu and Badescu (2011). This is the sunshine stability number SSSN, which is defined also as a random binary variable. In a simplified form the SSSN equation reads:

$$SSSN_t = \begin{cases} 1 & \text{if } \begin{cases} SSN_t > SSN_{t-1} \wedge SNN_1 = 0 \\ SSN_t < SSN_{t-1} \wedge SNN_1 = 1 \end{cases} \\ 0 & \text{otherwise} \end{cases} \quad (30)$$

Several comments about the definition of SSSN are useful. Depending of the initial value of SSN in a given time interval Δt, SSSN indicates the Sun's appearance/disappearance in the sky, while the entire sky may or may not be clear of clouds. The sum of $SSSN_t$ counts the number of times the Sun appears or disappears in/from the sky during Δt. The average value of SSSN over Δt, denoted $\overline{SSSN}$, ranges between zero (when the values of SSN are all 0 or 1, respectively, for all the moments t during Δt) and 1/2 (when the values of *SSN* change every two consecutive moments during

Δ*t*). The solar radiative regime is *fully stable* in the first case and *fully unstable* in the last case. Since $\overline{SSSN}$ is a measure of the frequency of the Sun appearance/disappearance in the sky, it quantifies straightforwardly the solar irradiance variability. This is illustrated in Figure 8, which displays the GHI and SSN values against local time.

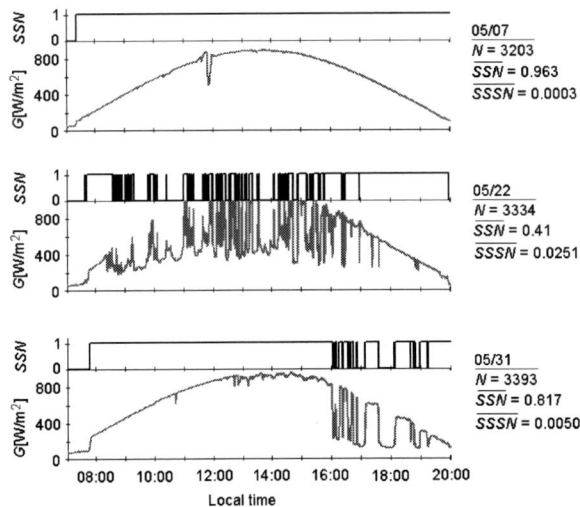

Figure 8. Global solar irradiance *G* and sunshine number *SSN* recorded on the Solar Platform of the West University of Timisoara, Romania during three days of May 2018. The number of measurements *N*, the daily average value of sunshine number SSN and the daily average value of sunshine stability number SSSN are indicated for each day.

The graphs are built with data recorded on the Solar Platform of the West University of Timisoara (Solar Platform 2019) in three days of May 2018. Each day experienced a very different variability in the state-of-the-sky. 05/07 was a beautiful day with cloudless sky. Conversely, a massive cloud-field passed over Timisoara in 05/22, generating a high-variability in the state-of-the-sky. 05/31 was a typical day for Timisoara, clear and stable in the morning and moderate variable afternoon due to the passing clouds. The variability in solar irradiance during these three days is well captured by the $\overline{SSSN}$ values: 0.0003 in 05/07 indicating a stable day, 0.0050 in

05/31 indicating a moderate variability in solar irradiance and 0.0281 in 05/22 indicating a high-variability in solar irradiance.

A comparison on the ability of different quantifiers to capture the peculiarities of the solar irradiance variability was reported by Blaga and Paulescu (2018). Six quantifiers, very different in their nature, were analyzed. These quantifiers are the stability index, the standard deviation of increments, the number of fronts, the integrated complementary cumulative distribution function of the increments, the sunshine stability number *SSSN* and the fractal dimension. The results show that *SSSN* is suitable for capturing the high-frequency fluctuations in solar irradiance.

Another use of SSN is in nowcasting GHI. The procedure was introduced by our team in 2014 (Paulescu et al. 2014). The resulted model was generically named the two-state model. Basically, it connects an empirical model for estimating GHI with a statistical model for nowcasting SSN. The the two-state model operates as follows: (1) If the Sun shines, the future value in the solar irradiance time series is estimated with an empirical clear-sky model; (2) If the Sun is covered by clouds, the clear-sky solar irradiance estimate is adjusted to the actual value of the cloud transmittance. The distinction between these two states is made by the forecasted value of SSN. Note that SSN is the sole quantity forecasted within the two-state model, which, considering the binary nature of SSN, is always a big challenge (e.g., Brabec et al. 2014). The general equation of the two-state model is (Paulescu et al. 2018):

$$\hat{G}_t = \begin{cases} c_{cs} \cdot G_{0,t} & \text{if } \hat{SSN}_t = 1 \\ \bar{\tau}_c \cdot G_{0,t} & \text{if } \hat{SSN}_t = 0 \end{cases} \tag{31}$$

where $\hat{G}_t$ is the forecasted value of GHI at time t. $G_{0,t}$ is the estimated solar irradiance under clear sky at time t, c_{cs} is a dynamic correction applied to the mean atmospheric transmittance encapsulated into the $G_{0,t}$ equation. $\bar{\tau}_c$ is an attenuation factor consistent with the cloud transmittance, that is applied to $G_{0,t}$. Both parameters c_{cs} and $\bar{\tau}_c$ are estimated simultaneously by a linear regression applied to data measured in a period Δt prior to the forecasting moment. Each day, the procedure of computing c_{cs} and $\bar{\tau}_c$ runs as follows. At the time moment when the Sun elevation angle reaches five degrees, c_{cs} and $\bar{\tau}_c$ are initialized: $c_{cs} = 1$, which means that $G_{0,t}$ is assumed to be an accurate estimator for the measured clear-sky solar irradiance and $\bar{\tau}_c = 0.29$, as the multiannual average of cloud transmittance. Further, c_{cs} and $\bar{\tau}_c$ are systematically updated before generating a forecast. In order to do that, the solar irradiance measurements recorded during a period Δt before the moment of generating the forecast are separated into two bins with respect to SSN. The measurements from the bin $SSN = 1$ are used for estimating the actual value of c_{cs} by the least squares method while the measurements from the bin $SSN = 0$ are used for estimating the actual value of $\bar{\tau}_c$. If a bin contains less than three measurements, the corresponding parameter, c_{cs} and $\bar{\tau}_c$, keeps the previous estimated value. Hence, the atmospheric transmittance and the cloud transmittance in Eq. (31) are always related to the most recent measurements.

Paulescu and Paulescu (2019) evaluated five statistical models for nowcasting GHI from different perspectives: forecast accuracy, forecasts precision, data-series granularity and variability in the data series. The five models are: random walk, moving average, exponential smoothing, autoregressive integrated moving average (ARIMA) and the two-state model. The results highlight the followings conclusions. Generally, the models performance decreases with increasing lead time. Different statistical indicators capture different particularities of the models. Looking at *nRMSE*, a very interesting and rather surprising conclusion is that the five models give similar forecasts. It is a consequence of large error encountered by all forecasts, which are strongly penalized by *RMSE*. Regardless of the lead time, the models are stratified in three levels of precision. The two-state model leads the hierarchy, followed by the random-walk model. The notable precision of the two-state model can be explained in terms of its structure, which allows a permanent adjustment to the actual atmospheric transmittance. This give a significative advantage to the two-state model when the percentage of the forecasts accurate to within a given tolerance interval is evaluated. The solar radiative regime variability has a strong influence on the performance of nowcasting solar irradiance: the performance of the models degrades as the instability increases. This major limitation of nowcasting accuracy is associated with the persistence property, i.e., the general tendency of a statistical model to extrapolate the current state in the future. A notable advantage of the two-state model was noted. This is related to its modularity, which facilitates to increase the forecasting accuracy by fine-tuning its components: the dynamic adjustment and the procedure for nowcasting SSN.

Forecasting DNI

A challenge standing against the growing share of solar electricity in the energy mix stems from the intrinsic nature of the solar energy which is stochastic fluctuating in time owing to irregular weather pattern. In order to reduce the costs of integrating the solar plants in the existing power grid, forecasting the energy generated by the solar plants is a key issue. A high-quality forecast will enable grid operators to plan the other capabilities (mainly gas power plants) to compensate for the solar plants power variations. Different models for forecasting DNI of different nature and complexity have been proposed over time. For example, Chu and Coimbra (2017) developed a k-nearest neighbor (kNN) ensemble model for generating the probability density function forecasts for intra-hour DNI. This probabilistic forecasting model uses diffuse irradiance measurements and cloud cover information as exogenous inputs. The results reported by authors show that the kNN model outperform the considered reference models in terms of all evaluation metrics and for all locations when the forecast horizon is greater than 5-min. The model demonstrated also notable-performances in predicting DNI ramps. Zhu et al. (2017) proposed an interesting model based on the wavelet transformation. The model consists of two components: (1) A semi-empirical model estimating DNI under clear-sky. The estimations are used for converting DNI series into a clear-sky index by removing the impact of the Sun's position. (2) A forecasting model which processes the clear-sky index using the wavelet technique. The series of the clear-sky index is decomposed in four sub-series with different frequencies. The authors constructed independent forecasting models for all sub-series. The reported result shows that the wavelet-based forecasting model achieves high-performance. Peruchena et al. (2017) observed the difficulties in accurately forecast DNI with NWP models at a high temporal resolution. The causes are related to the limitations in spatial and temporal resolution of NWP models, as well as the complexity in cloud microphysics and their radiative properties. However, NWP models provide the best forecast at a low temporal resolution (typically 1 to 3 hours in global scale modeling). Contrarily, statistical information derived from local measurements can provide high temporal forecasting of DNI but may not necessarily yield an explicit model for all the physical relationships involved. Based on these observations, Peruchena et al. (2017) propose a hybrid procedure for high-frequency forecasting of DNI that connects these two extremes.

An accurate forecast of DNI is important for achieving an accurate forecast of CSP output. Law et al. (2014) report a review focused on DNI forecast accuracy. Various models of different nature, based on numerical weather prediction (NWP), time series analysis, cloud motion vectors and hybrid methods, are considered. The results of the reviewed papers are summarized aiming to identify the best model for DNI forecasting. The authors note that DNI forecast information can be used by a CST plant operator to anticipate possible thermal shocks and sudden losses in electricity generation and make plans to reduce their negative impacts instead of reacting to them. The same authors evaluate the benefits of using 1-hour ahead forecasts to decide updated bids for a CST plant after making initial bids from 48-hours ahead forecasts (Law et al. 2016). The results of simulating a CST plant of 50 MWp with 7.5 hours of storage, operating in the Australian National Electricity Market for one year, showed that using

1-hour ahead forecasts increases substantially the financial value. The authors conclude that the operators of the CST plants can achieve higher financial value and reliability by using intra-day forecasts.

CONCLUSION

The first section of this chapter introduced two methods of measuring DNI at the ground level: (1) The pyrheliometer method, which is the most straightforward method for measuring DNI. The instrument should be permanently pointed towards the Sun. A two-axis Sun tracking mechanism is very often used for this purpose. Because of how expensive performant pyrheliometers and their maintenance are, a low spatial density of the stations equipped with such instruments is noted. (2) The pyranometric method, which is an indirect method. GHI and DHI are measured directly with pyranometers, then DNI is computed based on Eq. (1). The second section introduced the readers to modeling solar irradiance at ground level based on the clear-sky solar irradiance models. The physical basis of modeling DNI, along with three different classes of models (spectral codes, parametrical models and empirical models), were presented. Choosing a highly-performant parametric model is frequently limited by the availability of the parameters required for its running. Generally, the empirical models require at input only the geographical location and the time stamp, without any measured weather parameter. Being very simple, these models are widely preferred in various applications. However, the illustration of the models performance (see Sec. 3) shows that an empirical model should be chosen with care when the atmospheric conditions deviate considerable from normality (e.g., high atmospheric aerosol loading). The third section acknowledged the DNI forecasting. An accurate forecast of DNI is important for achieving an accurate forecast of the CSP output. As a non-traditional application, the role of DNI in defining what "the Sun shines in the sky" means was discussed. The sunshine number, a binary indicator stating whether the Sun is shining or not, converts the time series of DNI measurements into a binary series suitable for studying the variability in the state-of-the-sky or for improving the accuracy in nowcasting the global horizontal solar irradiance.

REFERENCES

Badescu V, Paulescu M (2011a) Statistical properties of the sunshine number illustrated with measurements from Timisoara (Romania). *Atmospheric Research* 101, 194-204.

Badescu V, Paulescu M (2011b) Autocorrelation properties of the sunshine number and sunshine stability number. *Meteorology and Atmospheric Physics* 112, 139-154.

Badescu V, Gueymard CA, Cheval S, Oprea C, Baciu M, Dumitrescu M, Iacobescu F, Milos I, Rada C (2013) Accuracy analysis for fifty-four clear-sky solar radiation models using routine hourly global irradiance measurements in Romania. *Renewable Energy* 55, 85-103.

Batlles FJ, Olmo FJ, Alados-Arboledas L (1995) On shadowband correction methods for diffuse irradiance measurements. *Solar Energy* 54, 105-114.

Bergman TL, Lavine AS, Incropera, FP Dewitt (2011) *Fundamentals of Heat and Mass Transfer,* 7th Ed, John Wiley & Sons.

Benkaciali S, Haddadi M, Khellaf (2018) Evaluation of direct solar irradiance from 18 broadband parametric models: Case of Algeria. *Renewable Energy* 125, 694-711.

Biga AJ, Rosa R (1979) Contribution to the study of the solar radiation climate of Lisbon. *Solar Energy* 23, 61–67.

Blaga R, Paulescu M (2018) Quantifiers for the solar irradiance variability: A new perspective. *Solar Energy* 174, 606-616.

Blanc P, Espinar B, Geuder N, Gueymard C, Meyer R, Pitz-Paal R, Reinhardt B, Renne D, Sengupta M, Wald L, Wilbert S (2014) Direct normal irradiance related definitions and applications: The circumsolar issue. *Solar Energy* 110, 561–577.

Brabec M, Paulescu M, Badescu V (2014) Generalized additive models for nowcasting cloud shading. *Solar Energy* 101, 272-282.

Calinoiu D, Stefu N, Boata R, Blaga R, Pop N, Paulescu E, Sabadus A, Paulescu M (2018) Parametric modeling: A simple and versatile route to solar irradiance. *Energy Conversion and Management* 164, 175-187.

Chauvin R, Nou J, Eynard J, Thil S, Grieu S (2018) A new approach to the real-time assessment and intraday forecasting of clear-sky direct normal irradiance. *Solar Energy* 167, 35–51.

Chu Y, Coimbra CFM (2017) Short-term probabilistic forecasts for Direct Normal Irradiance. *Renewable Energy* 101, 526-536.

Eissa Y, Marpu PR, Gherboudj I, Ghedira H, Ouarda TBMJ, Chiesa M (2013) Atificial neural network-based model for retrieval of the direct normal, diffuse horizontal and global horizontal irradiances using SEVIRI images. *Solar Energy* 89, 1–16.

Gueymard C (1995) *SMARTS2 – A Simple Model of the Atmospheric Radiative Transfer of Sunshine: Algorithms and performance assessment*. Florida Solar Energy Center Rep. FSEC-PF-270-95. https://www.nrel.gov/grid/solar-resource/smarts.html [Accessed July 2019].

Gueymard C (2018) A reevaluation of the solar constant based on a 42-year total solar irradiance time series and a reconciliation of spaceborne observations. *Solar Energy* 168, 2-9.

Gueymard C (2019) The SMARTS spectral irradiance model after 25 years: New developments and validation of reference spectra. *Solar Energy* 187, 233-253.

Gueymard C, Ruiz-Arias JA (2015) Validation of direct normal irradiance predictions under arid conditions: A review of radiative models and their turbidity-dependent performance. *Renewable and Sustainable Energy Reviews* 45, 379–396.

Habte A, Sengupta M, Andreas A, Wilcox S, Stoffel T (2015) Intercomparison of 51 radiometers for determining global horizontal irradiance and direct normal irradiance measurements. *Solar Energy* 133, 372–393.

Hottel HC (1976) A simple model for estimating the transmittance of direct solar radiation through clear atmosphere. *Solar Energy* 18, 129-139.

Kasten F, Young AT (1989) Revised optical air mass tables and approximation formula. *Applied Optics* 28, 4735-4738.

Kottek M, Grieser J, Beck C, Rudolf B, Rubel F (2006) *World Map of the Köppen-Geiger climate classification updated.* Meteorologische Zeitschrift 15, 259-263.

Kotti MC, Argiriou AA, Kazantzidis A (2014) Estimation of direct normal irradiance from measured global and corrected diffuse horizontal irradiance. *Energy* 70, 382-392.

Law EW, Prasad AA, Kay M, Taylor RA (2014) Direct normal irradiance forecasting and its application to concentrated solar thermal output forecasting – A review. *Solar Energy* 108, 287–307.

Law EW, Kay M, Taylor RA (2016) Evaluating the benefits of using short-term direct normal irradiance forecasts to operate a concentrated solar thermal plant. *Solar Energy* 140, 93–108.

Leckner B (1978) The spectral distribution of solar radiation at the earth's surface -elements of a model. *Solar Energy* 20, 143-150.

Nikitidou E, Kazantzidis A, Salamalikis (2014) The aerosol effect on direct normal irradiance in Europe under clear skies. *Rewable Energy* 68, 475-484.

Paulescu M, Schlett Z (2003) A simplified but accurate spectral solar irradiance model. *Theoretical and Applied Climatology* 75, 203-212.

Paulescu M, Badescu V (2011) New approach to measure the stability of the solar radiative regime. *Theoretical and Applied Climatology* 103, 459-470.

Paulescu M, Paulescu E (2019) Short-term forecasting of solar irradiance. *Renewable Energy* 143, 985-994.

Paulescu M, Paulescu E, Gravila P, Badescu V (2013) *Weather Modeling and Forecasting of PV Systems Operation*, Springer, London.

Paulescu M, Mares O, Paulescu E, Stefu N, Pacurar A, Calinoiu D, Gravila P, Pop N, Boata R (2014) Nowcasting solar irradiance using the sunshine number. *Energy Conversion and Management* 79, 690-697.

Paulescu M, Mares O, Dughir C, Paulescu E (2018) *Nowcasting the Output Power of PV Systems.* E3S Web Conf. 61, 00010.

Perez-Higueras PJ, Rodrigo P, Fernandez EF, Almonacid F, Hontoria L (2012) A simplified method for estimating direct normal solar irradiation from global horizontal irradiation useful for CPV applications. *Renewable and Sustainable Energy Reviews* 16, 5529–5534.

Peruchena CMF, Gaston M, Schroedter-Homscheidt M, Kosmale M, Martinez Marco I, García-Moya JA, Casado-Rubio JL (2017) Dynamic Paths: Towards high frequency direct normal irradiance forecasts. *Energy* 132, 315-323.

Porfirio ACS, Ceballos JC (2017) A method for estimating direct normal irradiation from GOES geostationary satellite imagery: Validation and application over Northeast Brazil. *Solar Energy* 155, 178–190.

Siren KE (1987) The shadow band correction for diffuse irradiation based on a two-component sky radiance model. *Solar Energy* 39, 433–438.

Solar Platform (2019) Solar Platform of the West University of Timisoara, Romania. http://solar.physics.uvt.ro/srms [Accessed in July 2019].

Vignola F, Michalsky J, Stoffel T (2017) *Solar and Infrared Radiation Measurements*. CRC Press, Boca Raton.

WMO (2008) *World Meteorological Organization. Guide to Meteorological Instruments and Methods of Observation*. WMO-No.8/2008 update in 2010. https://library.wmo.int/pmb_ged/wmo_8_en-2012.pdf [Accessed July 2019].

Yang K, Huang GW, Tamai N. A hybrid model for estimating global solar irradiance. *Solar Energy* 2001;70:13-22.

Younes S, Claywell R, Muneer T (2005) Quality control of solar radiation data: Present status and proposed new approaches. *Energy* 30, 1533–1549.

Zhu T, Wei H, Zhao X, Zhang C, Zhang K (2017) Clear-sky model for wavelet forecast of direct normal irradiance. *Renewable Energy* 104, 1-8.

In: Solar Irradiance
Editor: Daryl M. Welsh

ISBN: 978-1-53618-786-1
© 2020 Nova Science Publishers, Inc.

Chapter 3

POTENTIAL ASSESSMENT OF GLOBAL SOLAR RADIATION RESOURCES BASED ON EMPIRICAL AND ANN MODELS: ALGERIA AS A CASE STUDY

T. E. Boukelia[1,2,*], M. S. Mecibah[3], A. Ghellab[4] and A. Bouraoui[4]

[1]Mechanical Engineering Department, Jijel University, Jijel, Algeria
[2]Mechanical and Advanced Materials laboratory, Polytechnic School of Constantine, Constantine, Algeria
[3]Laboratory of Mechanics, Mechanical Engineering Department, Faculty of Technology Sciences, University of Brothers Mentouri, Constantine, Algeria
[4]Laboratory of Applied Energetics and Materials, Mechanical Engineering Department, Jijel University, Jijel, Algeria

---

[*] Corresponding Author's E-mail: taqy25000@hotmail.com.

Abstract

Solar energy is considered to be a clean, safe, and economic source of energy, and has the potential to be the most exploited energy source in the future. The amount of this type of energy, which refers to solar radiation available on the earth's surface, depends on different specifications between physical and astronomical, meteorological and geographical ones. These parameters or specifications can be expressed by and not limited to: distance between earth and sun, extraterrestrial radiation, different astronomical and solar angles, atmospheric transmittance, as well as sunshine duration, ambient temperature, humidity and cloudiness at the relevant locations.

To design and optimize any solar conversion system, the knowledge of accurate solar radiation data is extremely important. The best solar radiation data on the place of interest would be that measured at this specific location in a continuous and an accurate manner over a long term. However, for many developing countries, solar radiation measurements are not easily available due to financial or even technical limitations. On the other side, several spatial databases provide solar radiation values developed by different procedures, various spatial and temporal coverages, and different time intervals and space resolutions. However, these databases show different values of uncertainties due to the different approaches and the inputs used to generate them. Therefore, it is so important to elaborate solar radiation data based on high performance and simple models.

In this direction, the main aim of this chapter is to explore the potential of using simple empirical and ANN models to estimate global solar radiation on a horizontal surface. In this regard, Algeria was taken as a case study with eight different locations (Algiers, Oran, Batna, Constantine, Ghardaia, Bechar, Adrar, and Tamanrasset), and four statistical parameters were chosen to assess the performances of each model or approach. Furthermore, the results of these empirical correlations alongside with those of ANN model will be compared, in order to choose the best approach to generate global solar radiation databases in Algeria.

Introduction

Taking in consideration the vast sources of renewable energies, Algeria has established an ambitious strategy to use these sources to generate green power, and enhance their efficiencies. This program is focusing on a long-term policy started in 2011 and will end by 2030, to increase and extend the use of these inexhaustible resources, including wind, solar, as well as biomass and hydropower to support the national energy sector. Taking into account the enormous potential of solar energy in this Mediterranean country, the bulk of this program is devoted to systems based on photovoltaic and thermal solar technologies. More than two millions km^2 receives an annual sun exposure between 2000 kWh and 2900 kWh for each meter squared, and with a sunshine duration varies between 2650 h and 3500 h annually (Boukelia 2013). With these highly insolated large zones (exceed a total potential of 6 billion GWh annually), Algeria is one of countries with the highest solar radiation levels in the whole world (Stambouli et al. 2012).

This is from a side, on the other side, and before installing or erecting a solar conversion system, designing stage is extremely important to determine the techno-economic benefits of this type of systems as well as its feasibility. Consequently, the knowledge of accurate solar radiation data is highly required for engineers to select, design and predict the performances of these devices and systems. These data are generally based either on measurements provided by ground stations, spatial databases, or mathematical models. It is clear that the best data among these three methods, would be the first one which are measured at the location of interest continuously and accurately over a long period. However, these data are not easily available for many developing countries, especially Algeria, due to different financial or technical constraints. Therefore, there are still large areas in these countries without any observed solar radiation data. On the other hand, spatial data present different values of uncertainties, due to the developed approaches based on different inputs and processes, in addition to the lack of consistency in the measured data used by these spatial databases for interpolation, which are restricted to only a few ground stations. Hence, it is necessary to elaborate these data based on high performance and simple models. From this point of view, empirical models and artificial neural networks (ANN) can be good choices in the prediction and generation of solar radiation data. The regression models are well known by the scientific community as the most popular and employed approaches, because of their low computational cost, accessible inputs, and based on easy-to-use equations, however, and despite their complexities, ANN models present high accuracy in prediction compared to the first models.

In this direction, the main aim of this chapter is to explore the potential of using simple empirical and ANN models to estimate the monthly mean daily global solar radiation on a horizontal surface. In this regard, Algeria was taken as a case study through eight sites namely: Algiers, Oran, Batna, Constantine, Ghardaia, Bechar, as well as Adrar, and Tamanrasset, and four different statistical parameters were chosen to assess the performances of each model. Furthermore, the results of these empirical correlations alongside with those of ANN model will be compared, in order to choose the best databases for solar estimation in Algeria.

DATA AND METHODOLOGY

Solar Radiation Data

Solar radiation is produced by a nuclear fusion process generated inside the core of the sun, then it moves in the space in the form of electromagnetic waves, and reaches the earth's surface. The reached radiation involves two parts: extraterrestrial solar radiation above the atmosphere, and global one below the atmosphere which represents the first mentioned radiation attenuated when passing through the atmosphere layers. Therefore, solar radiation incident on earth's surface (global radiation) can further attenuate by different atmospheric processes, including absorption and scattering, which will change the diffuse and direct radiation ratio (Figure 1). As a consequence, the amount of solar radiation available on the earth's surface depends on many parameters, including and not limited to; distance between the sun and the earth atmospheric, transmittance, as well as latitude and cloudiness (Adaramola 2012).

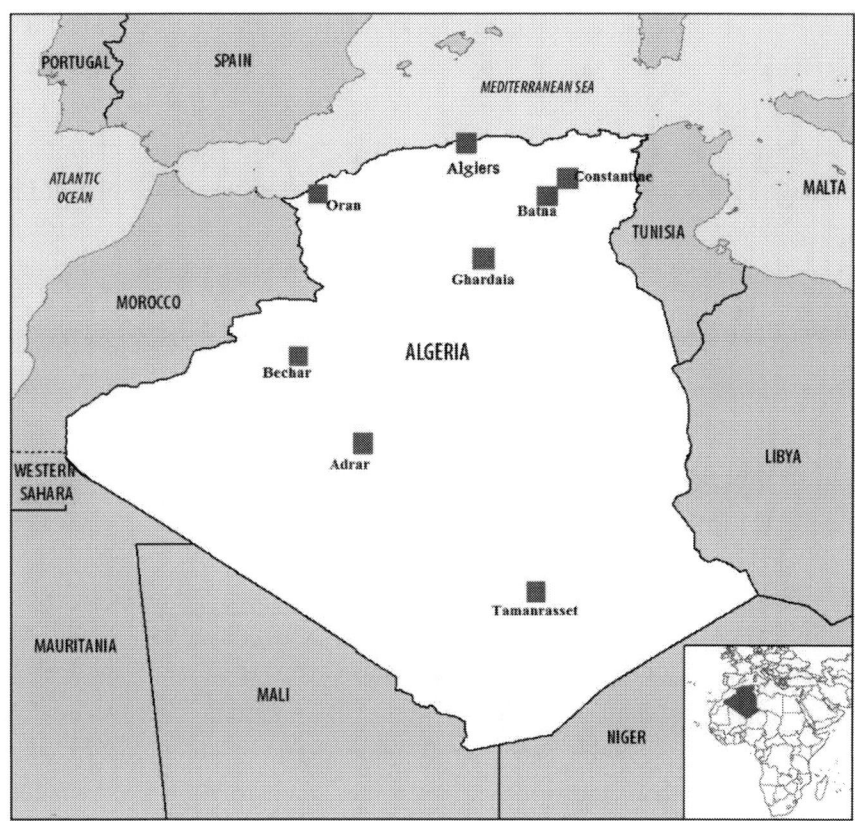

Figure 1. Locations of the defined stations on the map of Algeria.

Generally, and in many studies and works, it is necessary to know the amount of global radiation incident on a surface. Nevertheless, for designing and sizing of a solar conversion system, it is of great interest to know both the direct part of this radiation (for concentrating solar systems) (Boukelia 2015), as well as the diffuse part (for photovoltaic systems) (Kumar and Umanand 2005).

The observations of different meteorological and climatic parameters in Algeria are assured by the National Office of Meteorology (NOM) through its network composed of 81 long-term ground stations, however, only 7 stations record global solar radiation data. On the other side, the public Center dedicated to research and development of Renewable Energy Development (CDER) based in Algiers, alongside its annexes of Adrar and

Ghardaia provides the solar radiation data in these three locations (Mecibah et al. 2014). In this chapter, the considered generalized regression and ANN models were evaluated in eight Algerian locations with three different climates; Algiers and Oran with the Mediterranean climate in the northern part of the country; Constantine and Batna in the high plateaus zones and characterized by a semi-arid climate; Ghardaia, and Adrar, as well as Bechar, and Tamanrasset in the center and southern part of the country with a desert arid climate (Figure 1). Detailed geographic and the data period of these eight sites are presented in Table 1.

Regression and ANN Models

In order to predict the monthly mean data of daily global solar radiation in a simple way, two regression models (quadratic and cubic) in addition to ANN model were used. The most commonly used parameter for estimating global solar radiation based on regression models is sunshine duration, and the previous studies have assured that the models based on sunshine hours can provide more accurate prediction of solar radiation data than those based on other meteorological parameters (Almorox et al. 2011).

Table 1. Geographic and data records period of the studied locations

Sites	Latitude (N°)	Longitude (E°)	Altitude (m)	Sunshine duration (h)	Data series for establishing models	Data series for evaluating models
Algiers	36.43	3.15	25	7.79	1990-1992	2009-2011
Oran	35.38	-0.7	99	7.89	1990-1992	2009-2011
Constantine	36.28	6.61	694	7.45	2011-2012	2013
Batna	35.76	6.32	821	7.91	1990-1992	2009-2011
Ghardaia	32.36	3.81	450	8.86	2005-2006	2007-2009
Bechar	31.38	-2.15	806	9.78	2009-2010	2011
Adrar	27.82	-0.18	263.9	9.57	1990-1992	2009-2011
Tamanrasset	22.783	5.516	1377	8.92	1995-2005	2006-2011

Consequently, this study mainly is focused on sunshine based models. The other required parameters such as extraterrestrial radiation and maximum possible sunshine hours are worked out using the standard following relations (Duzen and Aydin 2012; Duffie and Beckman 1991; Bakirci 2009):

$$H_0 = \left(\frac{24\times 60}{\pi}\right) I_{sc} d_r [\cos(\varphi)\cos(\delta)\sin(\omega_s) + \omega_s \sin(\varphi)\sin(\delta)] \quad (1)$$

$$d_r = 1 + 0.033 \cos\left(\frac{2\pi}{365}J\right) \quad (2)$$

$$\delta = 0.4093 \sin\left[\frac{2\pi}{365}(248 + J)\right] \quad (3)$$

$$\omega_s = arc\cos[-\tan(\varphi)\tan(\delta)] \quad (4)$$

$$S_0 = \frac{2\omega_s}{15} \quad (5)$$

where d_r relative earth-sun distance, δ solar declination (rad), ω_s sunset hour angle (rad), φ latitude (rad), and J is the number of day in the year that taken from (Duffie and Beckman 1991). I_{sc} is the solar constant as 0.082 MJ/m^2/min (1367 W/m^2) (Duzen and Aydin 2012).

According to previous works, it could be reasonable to recommend the quadratic and cubic models based on sunshine records data (Mecibah et al. 2014, Boukelia et al. 2014), to use it as general models for estimating the global solar radiation on a horizontal surface over Algeria on the basis of their performances. While a, b, c, and d are the empirical coefficients which can be estimated using polynomial fitting of Matlab software package. The generalized forms of these two regression models were summarized in Table 2.

Table 2. Regression models used in this chapter

Models	Regression equations	Source
Quadratic	$\left(\dfrac{H}{H_0}\right) = a + b\left(\dfrac{S}{S_0}\right) + c\left(\dfrac{S}{S_0}\right)^2$	(Akinoglu and Ecevit 1990)
Cubic	$\left(\dfrac{H}{H_0}\right) = a + b\left(\dfrac{S}{S_0}\right) + c\left(\dfrac{S}{S_0}\right)^2 + d\left(\dfrac{S}{S_0}\right)^3$	(Bahel et al. 1987)

On the other side, and for the ANN model, the Levenberge Marguardt (LM) variant of the feed-forward back-propagation algorithm with a single hidden layer is presented in this study. Moreover, one of the key tasks in ANN prediction is the selection of the input variables, as well as the outputs. Latitude (φ), Altitude (Alt), number of the month (N_{month}), monthly mean daily sunshine duration (S), as well as mean minimum (T_{min}), and maximum (T_{min}) temperatures for each specific location are defined as the inputs, while monthly mean daily global solar radiation on a horizontal surface (H_g) is considered as the only output. In order to suit the consistency of this approach, the whole database is firstly normalized in the (0, 1) range, and then returned to original values after performing the simulations. The considered ANN model was carried out in two essential steps; training and testing. Thus, 70% of these data (96 × 7) were selected for training, while the rest were used for test step. On the other hand, the logarithmic sigmoid (Logsig) approach was used in our model, and it is presented as:

$$f(ze) = \frac{1}{1+e^{-ze}} \tag{6}$$

where ze is weighted sum, and presented in terms of bias (b), weight (w), and output (y) as:

$$ze_j = \sum_{i=1}^{n} w_{ij} y_i + b_j \tag{7}$$

In order to analyze the predicting ability of the proposed models, four statistical parameters have been used (COV, MPE, RMSE, R^2). These parameters were formulated in terms of output value (y_{output}), target value (y_{actual}), average of target ($\bar{y}_{actual}$) and pattern (n):

$$RMSE = \sqrt{\frac{1}{n}\sum_{i=1}^{n}(y_{output} - y_{actual})^2} \qquad (8)$$

$$MPE = \frac{1}{n}\sum_{i=1}^{n}\left(\frac{y_{output} - y_{actual}}{y_{output}}\right) \qquad (9)$$

$$COV = \frac{\sum_{i=1}^{n}(y_{output} - \bar{y}_{output})(y_{actual} - \bar{y}_{actual})}{n} \times 100 \qquad (10)$$

$$R^2 = \left[\frac{\sum_{i=1}^{n}(y_{output} - \bar{y}_{output})(y_{actual} - \bar{y}_{actual})}{\sqrt{\sum_{i=1}^{n}(y_{output} - \bar{y}_{output})^2 \sum_{i=1}^{n}(y_{actual} - \bar{y}_{actual})^2}}\right]^2 \qquad (11)$$

The ANN toolbox of Matlab has been used to perform the study in this chapter. In order to define the output accurately, 1000 epochs in each training process have been learned, while mean squared error (MSE) has been defined as an objective function.

RESULTS AND DISCUSSION

Before proceeding with the main results, and presenting the comparison between the prediction ability of the two methods (regression and ANN), a pre-analysis should be carried out to find the optimized size of the ANN hidden layer. The ANN model was presented with an increasing number of neurons, to define the optimal number that estimates the monthly mean daily global solar radiation accurately. The presented results of the statistical analysis of the testing part of the data using different sizes of the ANN were presented in Table 3.

Table 3. Statistical analysis of training data for ANN with different size of the hidden layer

Size of the hidden layer	COV	MPE	RMSE	R^2
2	8.0708	4.9203	1.8610	0.9490
8	1.2668	0.7772	0.2921	0.9987
10	0.9580	0.4972	0.2209	0.9993
12	0.0443	0.0238	0.0102	1.0000
14	0.0009	0.0006	0.0002	1.0000
20	0.0009	0.0006	0.0002	1.0000
30	0.0009	0.0006	0.0002	1.0000

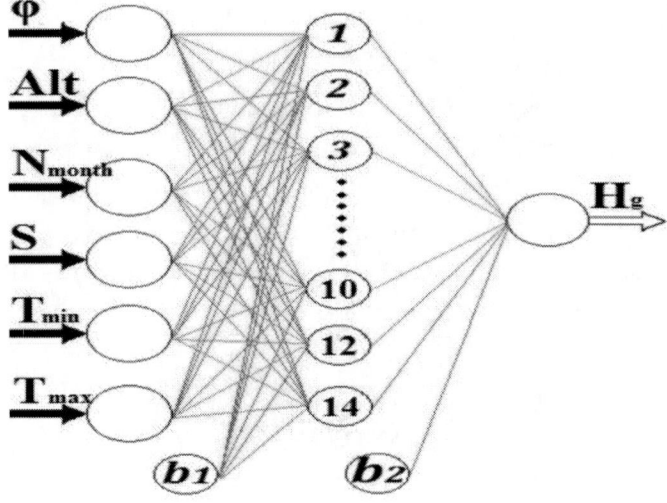

Figure 2. Architecture of the best ANN topology.

As it can be concluded from Table 3, the ANN model with different numbers of neurons can be used to predict the global solar radiation with an acceptable accuracy. However, the model with 14 neurons and onwards in a single hidden layer presents the highest accuracy. However, and in order to consider the easiness of the established models, the ANN with 14 neurons has been chosen as the optimized architecture to estimate the global solar radiation data (Figure 2).

Table 4. Statistical analysis of the two approaches of the eight locations

Location	Model	COV	MPE	RMSE	R^2
Algiers	Quadratic	10.7671	8.5222	1.7181	0.9522
	Cubic	10.8238	8.5599	1.7271	0.9521
	ANN	0.0012	0.0008	0.0002	1.0000
Oran	Quadratic	13.3724	10.6107	2.2227	0.9096
	Cubic	13.3944	10.6283	2.2264	0.9094
	ANN	0.0010	0.0007	0.0002	1.0000
Batna	Quadratic	5.7051	4.8286	0.9678	0.9823
	Cubic	5.7295	4.8666	0.9719	0.9822
	ANN	0.0011	0.0007	0.0002	1.0000
Constantine	Quadratic	8.5354	6.6385	1.3850	0.9861
	Cubic	8.5797	6.6737	1.3922	0.9860
	ANN	0.0009	0.0006	0.0002	1.0000
Ghardaia	Quadratic	16.3859	19.5843	3.4720	0.9848
	Cubic	16.3175	19.5009	3.4575	0.9848
	ANN	0.0006	0.0005	0.0002	1.0000
Bechar	Quadratic	15.8692	20.4071	3.4661	0.9704
	Cubic	15.8108	20.3266	3.4533	0.9704
	ANN	0.0007	0.0007	0.0002	1.0000
Adrar	Quadratic	14.5251	20.0814	3.0722	0.9917
	Cubic	14.4682	20.0016	3.0602	0.9917
	ANN	0.0010	0.0011	0.0003	1.0000
Tamanrasset	Quadratic	15.9126	30.4368	3.6440	0.9450
	Cubic	15.8487	30.3067	3.6294	0.9449
	ANN	0.0008	0.0010	0.0002	1.0000

In order to test the estimation capabilities by means of the two types of models; regression and ANN, and compare their performances with each other, a statistical analysis has been calculated based on the four above mentioned indicators. The results of this analysis have been outlined in Table 4, and the correlation coefficients of the proposed two generalized regression models (quadratic and cubic) are given as follows:

$$\left(\frac{H}{H_0}\right) = -0.034 + 1.2094\left(\frac{S}{S_0}\right) - 0.2783\left(\frac{S}{S_0}\right)^2 \qquad (12)$$

$$\left(\frac{H}{H_0}\right) = 0.3482 - 1.2422\left(\frac{S}{S_0}\right) + 4.7619\left(\frac{S}{S_0}\right)^2 - 3.3421\left(\frac{S}{S_0}\right)^3 \quad (13)$$

It is commonly familiar that the ideal values of statistical tests (RMSE, MPE, and COV) are 0 or close to 0, while the best values of R^2 should be 1 or close to 1. As clearly seen from this Table 4, good agreements can be obtained between the measured data of the monthly mean daily global solar radiation over the whole territory of Algeria, and those obtained by the two approaches (regression and ANN). However, the ANN model shows much higher performance compared to the two generalized regression models, the lowest values of RMSE, MPE, and COV and the highest values of R^2 at the same time are obtained by this approach. These values vary between the lowest prediction performances of COV = 0.0012, MPE = 0.0008, RMSE = 0.0002, R^2 = 1.0000 for Algiers, and the highest performances for Ghardaia region with COV = 0.0006, MPE = 0.0005, RMSE = 0.0002, R^2 = 1.0000. On the other hand, and as summarized in the same table, the quadratic correlation is the best model among the two established regression models for predicting the global solar radiation on a horizontal surface in the coastal regions (Algiers, and Oran) as well as in the high Plateaus zones (Batna and Constantine), which can reach the highest performances of prediction in Batna with values of 5.7051, 4.8286, 0.9678, and 0.9823 for COV, MPE, RMSE, and R^2 respectively. While in general, the cubic model reaches the highest performances in prediction for other regions with arid climate (Ghardaia, Bechar, Adrar, and Tamanrasset) with the values of COV= 14.4682, MPE = 20.0016, RMSE = 3.0602, R^2 = 0.9917 for Adrar. The reason for the higher prediction abilities of the ANN approach compared to the regression models can be justified that ANN behaves like an interpolation polynomial by forcing curves to pass from all experimental points (Gülüm 2018), and the extended number of inputs compared to the regression models.

Conclusion

The main objective of this chapter is to develop two approaches (on regression and ANN) for estimating global solar radiation on a horizontal surface, to predict the monthly mean daily data of this parameter, and compare their predictive ability with each other in order to choose the best and the easiest way to generate these data. The following results can be illustrated from this study:

- ANN model shows better performances in prediction of solar radiation data compared to regression models, as it gives the lowest values of RMSE, MPE, and COV and the highest values of R^2 at the same time. However, these regression models give the easiest and the simplest methodology.
- The quadratic correlation is the best approach among the two regression models in predicting the global solar radiation for the coastal and the high plateau regions. While the cubic model is the favor regression model for the southern regions with arid climate.

References

Adaramola, Muyiwa S. 2012. "Estimating global solar radiation using common meteorological data in Akure, Nigeria." *Renewable Energy* 47: 38-44.

Akinoğlu, B. G., and A. Ecevit. 1990. "A further comparison and discussion of sunshine-based models to estimate global solar radiation." *Energy* 15.10: 865-72.

Almorox, J., C. Hontoria, and M. Benito. 2011. "Models for obtaining daily global solar radiation with measured air temperature data in Madrid (Spain)." *Applied Energy* 88.5: 1703-09.

Bahel, V., H. Bakhsh, and Ra Srinivasan. 1987. "A correlation for estimation of global solar radiation." *Energy* 12.2: 131-5.

Bakirci, Kadir. 2009. "Correlations for estimation of daily global solar radiation with hours of bright sunshine in Turkey." *Energy* 34.4: 485-501.

Boukelia, Taqiy Eddine, and Mohamed Salah Mecibah. 2015. "Estimation of direct solar irradiance intercepted by a solar concentrator in different modes of tracking (case study: Algeria)." *International Journal of Ambient Energy* 36.6: 301-8.

Boukelia, Taqiy eddine, and Mohamed-Salah Mecibah. 2013. "Parabolic trough solar thermal power plant: Potential, and projects development in Algeria." *Renewable and Sustainable Energy Reviews* 21: 288-97.

Boukelia, Taqiy Eddine, Mohamed-Salah Mecibah, and Imad Eddine Meriche. 2014. "General models for estimation of the monthly mean daily diffuse solar radiation (Case study: Algeria)." *Energy Conversion and Management* 81: 211-9.

Duffie, John A., and William A. Beckman. 2013. *Solar engineering of thermal processes*. 2nd ed. New York: John Wiley & Sons.

Duzen, Hacer, and Harun Aydin. 2012. "Sunshine-based estimation of global solar radiation on horizontal surface at Lake Van region (Turkey)." *Energy Conversion and Management* 58: 35-46.

Gülüm, Mert, Funda Kutlu Onay, and Atilla Bilgin. 2018. "Comparison of viscosity prediction capabilities of regression models and artificial neural networks." *Energy* 161: 361-9.

Kumar, Ravinder, and L. Umanand. 2015. "Estimation of global radiation using clearness index model for sizing photovoltaic system." *Renewable Energy* 30.15: 2221-33.

Mecibah, Mohamed Salah, et al. 2014. "Introducing the best model for estimation the monthly mean daily global solar radiation on a horizontal surface (Case study: Algeria)." *Renewable and Sustainable Energy Reviews* 36: 194-202.

Stambouli, A. Boudghene, et al. 2012. "A review on the renewable energy development in Algeria: Current perspective, energy scenario and sustainability issues." *Renewable and Sustainable Energy Reviews* 16.7: 4445-60.

INDEX

A

accuracy, 39, 82, 84, 85, 91, 100, 101, 102, 103, 104, 106, 111, 113, 115, 122, 128
aerosols, 92, 96, 98, 99, 100
air temperature, 132
Algeria, iv, v, vii, x, 104, 105, 115, 119, 120, 121, 122, 123, 125, 130, 132, 133
algorithm, viii, ix, 2, 3, 9, 10, 11, 12, 13, 16, 19, 20, 21, 22, 25, 28, 29, 30, 31, 32, 35, 37, 38, 39, 40, 46, 47, 48, 50, 51, 52, 53, 54, 56, 57, 58, 59, 60, 61, 62, 63, 64, 65, 66, 67, 68, 69, 70, 71, 72, 126
artificial intelligence, vii, 1, 9, 71, 72
atmosphere, 82, 91, 92, 93, 94, 95, 96, 98, 106, 116, 122
atmospheric pressure, 97
atmospheric transmittance, 96

B

basic research, 71, 77, 78
batteries, 7, 8, 44, 45, 46, 47
benefits, 19, 36, 61, 67, 70, 113, 117, 121

Boltzmann constant, 8, 33
broadband, 87, 88, 101, 102, 115

C

cell surface, 83
classification, 28, 34, 89, 104, 105, 117
climate, 103, 104, 105, 106, 115, 117, 124, 130, 131
coastal region, 130
comparative analysis, 103
computing, vii, x, 82, 111
conductance, ix, 2, 11, 31, 69
configuration, ix, 2, 12, 46, 50, 70
convergence, 10, 11, 14, 16, 20, 22, 28, 30, 31, 47, 53, 56, 60, 63, 64, 66, 68, 69
correlation, 129, 130, 131, 132
correlation coefficient, 129
correlations, x, 120, 122
cost, 3, 7, 8, 44, 45, 46, 47, 50, 52, 70, 83, 122
cumulative distribution function, 110

Index

D

database, 10, 13, 16, 22, 40, 47, 68, 126
dependent variable, 41
developing countries, x, 120, 121
distribution, 47, 52, 71, 85, 98, 99, 117

E

electricity, 9, 83, 112, 113
electromagnetic, 84, 122
electromagnetic waves, 122
energy, vii, x, 1, 9, 37, 51, 52, 53, 62, 67, 70, 71, 72, 82, 83, 85, 86, 87, 91, 93, 99, 112, 120, 121, 133
engineering, 96, 101, 132
environment, 32, 53, 55, 68, 103
environmental conditions, 32
evaporation, 8, 9, 57, 58, 59, 60, 63
extinction, 15, 93, 94, 96, 97, 101
extraterrestrial radiation variation, 92

F

financial, x, 114, 120, 121
fitness, 18, 19, 20, 21, 23, 24, 30, 35, 36, 58, 59
fluctuations, 15, 31, 37, 110
forecasting, vii, viii, 1, 2, 3, 9, 10, 11, 13, 16, 19, 20, 21, 27, 29, 30, 67, 68, 69, 73, 79, 80, 107, 111, 112, 113, 114, 116, 117
forecasting model, viii, 2, 10, 11, 13, 30, 69, 112

G

global scale, 113

global solar radiation, iv, v, vii, x, 119, 120, 122, 123, 124, 125, 126, 127, 128, 130, 131, 132, 133

H

historical data, 8, 9, 16, 22
hourly photovoltaic system power forecasting, 3
humidity, x, 101, 120

I

information processing, 72, 76, 78
irradiance diffuse, 82
irradiance empiric model, 102
irradiation, 36, 39, 50, 53, 54, 61, 64, 66, 67, 69, 70, 90, 118
iteration, 5, 10, 11, 13, 14, 16, 20, 22, 23, 24, 25, 26, 30, 54, 57, 58, 60, 67, 69, 71

L

life cycle, 45, 47, 50, 70
linear dependence, 86

M

machine learning, vii, 1, 9
maximum power point tracking, ix, 2, 3, 12, 31, 53, 54, 66, 69, 71
Mediterranean climate, 124
membership, viii, 2, 9, 28, 29, 32, 34, 35, 67
models, vii, viii, ix, x, 1, 9, 10, 12, 19, 32, 37, 40, 67, 68, 69, 71, 81, 83, 84, 90, 91, 94, 95, 96, 101, 102, 103, 104, 105, 106, 111, 112, 113, 114, 115, 116, 120, 121, 122, 124, 125, 126, 128, 129, 130, 131, 132

modified fuzzy neural net, iv, v, vii, viii, 1, 2, 3, 6, 9, 10, 11, 12, 16, 19, 21, 29, 31, 32, 36, 37, 39, 40, 46, 47, 50, 52, 67, 68, 69, 70, 71, 72, 78
modules, 6, 7, 8, 14, 40, 41, 43, 44, 45, 46, 47, 55, 62
motivation, viii, 1, 9, 11, 12, 21, 32, 54, 67

N

neural network, vii, 12, 19, 21, 28, 32, 34, 35, 39, 75, 78, 116, 121, 132
neural networks, 12, 21, 28, 32, 34, 39, 75, 78, 121, 132
neurons, 19, 28, 30, 34, 36, 127, 128
nodes, 11, 22, 23, 27, 34, 69

O

optimization, 10, 11, 12, 21, 28, 35, 40, 45, 46, 47, 48, 50, 51, 53, 54, 56, 59, 61, 65, 66, 68, 70, 75
ozone, 96, 97, 98, 99, 100, 102

P

parametric model, ix, 82, 91, 101, 103, 106, 114, 115, 116
photovoltaic system, iv, vii, 1, 2, 3, 6, 9, 10, 11, 12, 13, 32, 36, 39, 40, 44, 46, 50, 52, 53, 67, 68, 69, 70, 71, 72, 73, 74, 78, 79, 80, 123, 132
photovoltaic system's control, 3, 32, 79
plants, 52, 62, 70, 83, 107, 112, 114
power plants, 52, 82, 112
probability, 40, 60, 112
probability density function, 112

R

radiation, vii, ix, x, 6, 14, 33, 40, 41, 81, 82, 83, 84, 85, 87, 88, 90, 91, 92, 93, 94, 95, 96, 97, 98, 104, 115, 116, 117, 118, 120, 121, 122, 123, 124, 125, 126, 127, 128, 130, 131, 132, 133
regions of the world, 53
regression, 111, 121, 124, 125, 126, 127, 129, 130, 131, 132
regression model, 121, 124, 125, 129, 130, 131, 132
renewable energy, 53, 133
requirements, 8, 47, 86
resolution, 84, 90, 100, 104, 113
resources, 53, 84, 85, 121
response, 12, 31, 37, 39, 56, 67, 69, 86
response time, 12, 31, 37, 39, 56, 69, 86

S

scattering, 82, 92, 94, 96, 97, 99, 102, 122
simulation, viii, ix, 2, 11, 12, 14, 22, 29, 36, 37, 39, 44, 47, 53, 54, 61, 65, 66, 67, 68, 69
simulations, 36, 61, 70, 126
software, 19, 29, 50, 51, 125
solar irradiance, iv, vii, 1, 2, 4, 5, 8, 9, 11, 12, 15, 31, 32, 34, 36, 37, 38, 44, 47, 53, 55, 57, 61, 64, 65, 66, 67, 75, 79, 82, 83, 84, 87, 89, 90, 91, 92, 96, 97, 99, 100, 101, 102, 104, 106, 107, 108, 109, 110, 112, 114, 115, 116, 117, 118, 132
solar system, vii, ix, 72, 73, 82, 107, 123
solution, 9, 10, 11, 12, 14, 16, 19, 21, 22, 27, 28, 35, 40, 46, 52, 56, 69, 70, 86
statistical, iv, vii, x, 19, 29, 36, 104, 107, 110, 111, 113, 115, 120, 122, 126, 127, 128, 129, 130
storage, 7, 40, 44, 70, 113
structure, 16, 22, 27, 112

sunshine number, iv, vii, x, 82, 107, 109, 114, 115, 117

T

temperature, viii, x, 2, 8, 10, 11, 13, 16, 21, 22, 27, 29, 30, 33, 40, 47, 50, 53, 55, 68, 69, 70, 85, 86, 87, 94, 101, 120
testing, 90, 105, 106, 126, 127
time resolution, 103
time series, 40, 91, 110, 113, 114, 116
training, viii, 2, 11, 14, 19, 20, 22, 25, 29, 34, 36, 68, 69, 126, 127, 128
transactions, 10, 13, 68
transformation, 113
two-state model, 110, 111

U

uniform, ix, 2, 12, 53, 54, 55, 61, 64, 65, 66, 67, 94

V

variables, 6, 41, 86, 102, 126
variations, 11, 21, 31, 68, 87, 112
velocity, 8, 47, 53, 56, 57, 60

W

water vapor, 88, 96, 97, 102
wavelengths, 87, 98, 100